PROBLEM
SOLVING
in
MATHEMATICS

Cover design by Bev and Charles Dana

This work was developed under an ESEA Title IVC grant from the Oregon Department of Education, Office of Policy and Program Development. The content, however, does not necessarily reflect the position or policy of the Oregon Department of Education and no official endorsement of these materials should be inferred.

Distribution for this work was arranged by LINC Resources, Inc.

ISBN 0-86651-182-2

Order Number DS01408

11 12 13 14 15 16 17 18 19 20 -ML- 99 98 97

DALE
SEYMOUR
PUBLICATIONS
P.O. BOX 10888
PALO ALTO, CA 94303

PROBLEM SOLVING IN MATHEMATICS

PROJECT STAFF

DIRECTOR: OSCAR SCHAAF, UNIVERSITY OF OREGON
ASSOCIATE DIRECTOR: RICHARD BRANNAN, LANE EDUCATION SERVICE DISTRICT

WRITERS: RICHARD BRANNAN
                         MARYANN DEBRICK
                         JUDITH JOHNSON
                         GLENDA KIMERLING
                         SCOTT McFADDEN
                         JILL McKENNEY
                         OSCAR SCHAAF
                         MARY ANN TODD

PRODUCTION: MEREDITH SCHAAF
                         BARBARA STOEFFLER

EVALUATION: HENRY DIZNEY
                         ARTHUR MITTMAN
                         JAMES ELLIOTT
                         LESLIE MAYES
                         ALISTAIR PEACOCK

PROJECT GRADUATE   FRANK DEBRICK
  STUDENTS: MAX GILLETT
                         KEN JENSEN
                         PATTY KINCAID
                         CARTER McCONNELL
                         TOM STONE

ACKNOWLEDGEMENTS:

TITLE IV-C LIAISON: Ray Talbert
         Charles Nelson

         <u>Monitoring Team</u>

         Charles Barker
         Ron Clawson
         Jeri Dickerson
         Anthony Fernandez
         Richard Olson
         Ralph Parrish
         Fred Rugh
         Alton Smedstad

ADVISORY COMMITTEE: Mary Grace Kantowski University of Florida
         John LeBlanc   Indiana University
         Richard Lesh   Northwestern University
         Edwin McClintock   Florida International University
         Len Pikaart   Ohio University
         Kenneth Vos   The College of St. Catherine

A special thanks is due to the many teachers, schools, and districts within the state of Oregon that have participated in the development and evaluation of the project materials. A list would be lengthy and certainly someone's name would inadvertently be omitted. Those persons involved have the project's heartfelt thanks for an impossible job well done.

The following projects and/or persons are thanked for their willingness to share pupil materials, evaluation materials, and other ideas.

  Don Fineran, Mathematics Consulant, Oregon Department of Education
  Frank Lester, Indiana University
  Steve Meiring, Mathematics Consultant, Ohio Department of Education
  Harold Schoen, University of Iowa
  Iowa Problem Solving Project, Earl Ockenga, Manager
  Math Lab Curriculum for Junior High, Dan Dolan, Director
  Mathematical Problem Solving Project, John LeBlanc, Director

CONTENTS

INTRODUCTION

What is PSM?

PROBLEM SOLVING IN MATHEMATICS is a program of problem-solving lessons and teaching techniques for grades 4–8 and (9) algebra. Each grade-level book contains approximately 80 lessons and a teacher's commentary with teaching suggestions and answer key for each lesson. *Problem Solving in Mathematics* is not intended to be a complete mathematics program by itself. Neither is it supplementary in the sense of being extra credit or to be done on special days. Rather, it is designed to be integrated into the regular mathematics program. Many of the problem-solving activities fit into the usual topics of whole numbers, fractions, decimals, percents, or equation solving. Each book begins with lessons that teach several problem-solving skills. Drill and practice, grade-level topics, and challenge activities using these problem-solving skills complete the book.

PROBLEM SOLVING IN MATHEMATICS is designed for use with all pupils in grades 4–8 and (9) algebra. At-grade-level pupils will be able to do the activities as they are. More advanced pupils may solve the problems and then extend their learning by using new data or creating new problems of a similar nature. Low achievers, often identified as such only because they haven't reached certain computational levels, should be able to do the work in PSM with minor modifications. The teacher may wish to work with these pupils at a slower pace using more explanations and presenting the material in smaller doses.

[Additional problems appropriate for low achievers are contained in the *Alternative Problem Solving in Mathematics* book. Many of the activities in that book are similar to those in the regular books except that the math computation and length of time needed for completion are scaled down. The activities are generally appropriate for pupils in grades 4–6.]

Why Teach Problem Solving?

Problem solving is an ability people need throughout life. Pupils have many problems with varying degrees of complexity. Problems arise as they attempt to understand concepts, see relationships, acquire skills, and get along with their peers, parents, and teachers. Adults have problems, many of which are associated with making a living, coping with the energy crisis, living in a nation with peoples from different cultural backgrounds, and preserving the environment. Since problems are so central to living, educators need to be concerned about the growth their pupils make in tackling problems.

What Is a Problem?

MACHINE HOOK-UPS

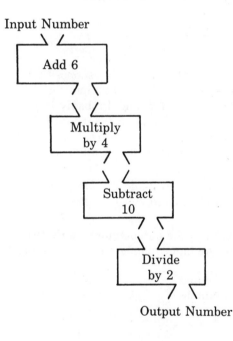

Input Number

Output Number

It is highly recommended that teachers intending to use *Problem Solving in Mathematics* receive training in implementing the program. The *In-Service Guide* contains much of this valuable material. In addition, in-service audio cassette tapes are available. These provide indepth guidance on using the PSM grade-level books and an overall explanation of how to implement the whole program. The tapes are available for loan upon request. Please contact Dale Seymour Publications, Box 10888, Palo Alto, CA 94303 for further information about the tapes and other possible in-service opportunities.

	Input Number	Output Number
a.	4	
b.	8	
c.	12	
d.		39
e.		47
f.		61

Suppose a 6th grader were asked to fill in the missing output blanks for a, b, and c in the table. Would this be a problem for him? Probably not, since all he would need to do is to follow the directions. Suppose a 2nd-year algebra student were asked to fill in the missing input blank for d. Would this be a problem? Probably not, since she would write the suggested equation,

$$\frac{4(x+6)-10}{2} = 39$$

and then solve it for the input. Now suppose the 6th grader were asked to fill in the input for d, would this be a problem for him? Probably it *would* be. He has no directions for getting the answer. However, if he has the desire, it is within his power to find the answer. What might he do? Here are some possibilities:

1. He might make *guesses*, do *checking*, and then make refinements until he gets the answer.
2. He might fill in the output numbers that correspond to the input numbers for a, b, and c.

	Input Number	Output Number
a.	4	
b.	8	
c.	12	
d.		39

and then observe this pattern:
For an increase of 4 for the input, the output is increased by 8.
Such an observation should lead quickly to the required input of 16.

3. He might start with the output and *work backwards* through the machine hook-up using the inverse (or opposite) operations.

For this pupil, there was no "ready-made" way for him to find the answer, but most motivated 6th-grade pupils would find a way.

A *problem*, then, is a situation in which an individual or group accepts the challenge of performing a task for which there is no immediately obvious way to determine a solution. Frequently, the problem can be approached in many ways. Occasionally, the resulting investigations are nonproductive. Sometimes they are so productive as to lead to many different solutions or suggest more problems than they solve.

What Does Problem Solving Involve?

Problem solving requires the use of many *skills*. Usually these skills need to be used in certain combinations before a problem is solved. A combination of skills used in working toward the solution of a problem can be referred to as a *strategy*. A successful strategy requires the individual or group to generate the information needed for solving the problem. A considerable amount of creativity can be involved in generating this information.

What Problem-Solving Skills Are Used in PSM?

Skills are the building blocks used in solving a problem. The pupil materials in the PSM book afford many opportunities to emphasize problem-solving skills. A listing of these skills is given below.

THE PSM CLASSIFIED LIST
OF PROBLEM-SOLVING SKILLS

A. Problem Discovery, Formulation
1. State the problem in your own words.
2. Clarify the problem through careful reading and by asking questions.
3. Visualize an object from its drawing or description.
4. Follow written and/or oral directions.

B. Seeking Information

5. Collect data needed to solve the problem.
6. Share data and results with other persons.
7. Listen to persons who have relevant knowledge and experiences to share.
8. Search printed matter for needed information.
9. Make necessary measurements for obtaining a solution.
10. Record solution possibilities or attempts.
11. Recall and list related information and knowledge.

C. Analyzing Information

12. Eliminate extraneous information.
13. Find likenesses and differences and make comparisons.
14. Classify objects or concepts.
15. Make and use a drawing or model.
16. Make and/or use a systematic list or table.
17. Make and/or use a graph.
18. Look for patterns and/or properties.
19. Use mathematical symbols to describe situations.
20. Break a problem into manageable parts.

D. Solve—Putting It Together—Synthesis

21. Make predictions, conjectures, and/or generalizations based upon data.
22. Make decisions based upon data.
23. Make necessary computations needed for the solution.
24. Determine limits and/or eliminate possibilities.
25. Make reasonable estimates.
26. Guess, check, and refine.
27. Solve an easier but related problem. Study solution process for clues.
28. Change a problem into one you can solve. (Simplify the problem.)
29. Satisfy one condition at a time.
30. Look at problem situation from different points of view.
31. Reason from what you already know. (Deduce.)
32. Work backwards.
33. Check calculated answers by making approximations.
34. Detect and correct errors.
35. Make necessary measurements for checking a solution.

36. Identify problem situation in which a solution is not possible.
37. Revise the conditions of a problem so a solution is possible.

E. Looking Back—Consolidating Gains

38. Explain how you solved a problem.
39. Make explanations based upon data.
40. Solve a problem using a different method.
41. Find another answer when more than one is possible.
42. Double check solutions by using some formal reasoning method (mathematical proof).
43. Study the solution process.
44. Find or invent other problems which can be solved by certain solution procedures.
45. Generalize a problem solution so as to include other solutions.

F. Looking Ahead—Formulating New Problems

46. Create new problems by varying a given one.

What Are Some Examples of Problem-Solving Strategies?

Since strategies are a combination of skills, a listing (if it were possible) would be even more cumbersome than the list of skills. Examples of some strategies that might be used in the "Machine Hook-Ups" problem follow:

Strategy 1. *Guess* the input; *check* by computing the output number for your guess; if guess does not give the desired output, note the direction of error; *refine* the guess; compute; continue making refinements until the correct output results.

Strategy 2. *Observe* the *patterns* suggested by the input and output numbers for the *a*, *b*, *c* entries in the table; *predict* additional output and input numbers by extending both patterns; *check* the predicted input for the *d* entry by computing.

Strategy 3. *Study* the operations suggested in the machine hook-up; *work backwards* through the machine *using previous knowledge* about inverse operations.

An awareness of the strategies being used to solve a problem is probably the most important step in the development of a pupil's problem-solving abilities.

What is the Instructional Approach Used in PSM?

The content objectives of the lessons are similar to those of most textbooks. The difference is in the approach used. First, a wider variety of problem-solving skills is emphasized in the materials than in most texts. Second, different styles of teaching such as direct instruction, guided discovery, laboratory work, small-group discussions, nondirective instruction, and individual work all have a role to play in problem-solving instruction.

Most texts employ direct instruction almost exclusively, whereas similar lessons in PSM are patterned after a guided discovery approach. Also, an attempt is made in the materials to use intuitive approaches extensively before teaching formal algorithms. Each of the following is an integral part of the instructional approach to problem solving.

A. TEACH PROBLEM-SOLVING SKILLS DIRECTLY

Problem-solving skills such as "follow directions," "listen," and "correct errors" are skills teachers expect pupils to master. Yet, such skills as "guess and check," "make a systematic list," "look for a pattern," or "change a problem into one you can solve" are seldom made the object of direct instruction. These skills, as well as many more, need emphasis. Detailed examples for teaching these skills early in the school year are given in the commentaries to the *Getting Started* activities.

B. INCORPORATE A PROBLEM-SOLVING APPROACH WHEN TEACHING TOPICS IN THE COURSE OF STUDY

Drill and practice activities. Each PSM book includes many pages of drill and practice at the problem-solving level. These pages, along with the *Getting Started* section, are easy for pupils and teachers to get into and should be started early in the school year.

Laboratory activities and investigations involving mathematical applications and readiness activities. Readiness activities from such mathematical strands as geometry, number theory, and probability are included in each book. For example, area explorations are used in grade 4 as the initial stage in the teaching of the multiplication and division algorithms and fraction concepts.

Teaching mathematical concepts, generalizations, and processes. Each book includes two or more sections on grade-level content topics. For the most part, these topics are developmental in nature and usually need to be supplemented with practice pages selected from a textbook.

C. PROVIDE MANY OPPORTUNITIES FOR PUPILS TO USE THEIR OWN PROBLEM-SOLVING STRATEGIES

One section of each book includes a collection of challenge activities which provide opportunities for emphasizing problem-solving strategies. Generally, instruction should be nondirective, but at times suggestions may need to be given. If possible, these suggestions should be made in the form of alternatives to be explored rather than hints to be followed.

D. CREATE A CLASSROOM ATMOSPHERE IN WHICH OPENNESS AND CREATIVITY CAN OCCUR

Such a classroom climate should develop if the considerations mentioned in A, B, and C are followed. Some specific suggestions to keep in mind as the materials are used are:

- Set an example by solving problems and by sharing these experiences with the pupils.
- Reduce anxiety by encouraging communication and cooperation. On frequent occasions problems might be investigated using a cooperative mode of instruction along with brainstorming sessions.
- Encourage pupils in their efforts to solve a problem by indicating that their strategies are worth trying and by providing them with sufficient time to investigate the problem; stress the value of the procedures pupils use.
- Use pupils' ideas (including their mistakes) in solving problems and developing lessons.
- Ask probing questions which make use of words and phrases such as
  I wonder if . . . .
  Do you suppose that . . . .
  What happens if . . . .
  How could we find out . . . .
  Is it possible that . . . .
- Reinforce the asking of probing questions by pupils as they search for increased understanding. Pupils seldom are skilled at seeking probing questions but they can be taught to do so. If instruction is successful, questions of the type, "What should I do now?," will be addressed to themselves rather than to the teacher.

What Are the Parts of Each PSM Book?

Grade 4	Grade 5	Grade 6	Grade 7	Grade 8	Grade 9
Getting Started	Getting Started	Getting Started	Getting Started	Getting Started	Getting Started
Place Value Drill and Practice	Whole Number Drill and Practice	Drill and Practice	Drill and Practice- Whole Numbers	Drill and Practice	Algebraic Concepts and Patterns
Whole Number Drill and Practice	Story Problems	Story Problems	Drill and Practice- Fractions	Variation	Algebraic Explanations
Multiplication and Division Concepts	Fractions	Fractions	Drill and Practice- Decimals	Integer Sense	Equation Solving
Fraction Concepts	Geometry	Geometry	Percent Sense	Equation Solving	Word Problems
Two-digit Multiplication	Decimals	Decimals	Factors, Multiples, and Primes	Protractor Experiments	Binomials
Geometry	Probability	Probability	Measurement-Volume, Area, Perimeter	Investigations in Geometry	Graphs and Equations
Rectangles and Division	Estimation with Calculators	Challenges	Probability	Calculator	Graph Investigations
Challenges	Challenges		Challenges	Percent Estimation	Systems of Linear Equations
				Probability	Challenges
				Challenges	

Notice that the above chart is only a scope of PSM—not a scope and sequence. In general, no sequence of topics is suggested with the exceptions that *Getting Started* activities must come early in the school year and *Challenge* activities are usually deferred until later in the year.

Getting Started Several problem-solving skills are presented in the *Getting Started* section of each grade level. Hopefully, by concentrating on these skills during the first few weeks of school pupils will have confidence in applying them to problems that occur later on. In presenting these skills, a direct mode of instruction is recommended. Since the emphasis needs to be on the problem-solving skill used to find the solution, about ten to twelve minutes per day are needed to present a problem.

Drill And Practice No sequence is implied by the order of activities included in these sections. They can be used throughout the year but are especially appropriate near the beginning of the year when the initial chapters in the textbook emphasize review. Most of the activities are not intended to develop any particular concept. Rather, they are drill and practice lessons with a problem-solving flavor.

Challenges Fifteen or more challenge problems are included in each book. In general, these should be used only after *Getting Started* activities have been completed and pupils have had some successful problem-solving experiences.

Many of the other sections in PSM are intended to focus on particular grade-level content. The purpose is to provide intuitive background for certain topics. A more extensive textbook treatment usually will need to follow the intuitive development.

Teacher Commentaries Each section of a PSM book has an overview teacher commentary. The overview commentary usually includes some philosophy and some suggestions as to how the activities within the section should be used. Also, every pupil page in PSM has a teacher commentary on the back of the lesson. Included here are mathematics teaching objectives, problem-solving skills pupils might use, materials needed, comments and suggestions, and answers.

How Often Should Instruction Be Focused on Problem Solving?

Some class time should be given to problem solving nearly every day. On some days an entire class period might be spent on problem-solving activities; on others, only 8 to 10 minutes. Not all the activities need to be selected from PSM. Your textbook may contain ideas. Certainly you can create some of your own. Many companies now have published excellent materials which can be used as sources for problem-solving ideas. Frequently, short periods of time should be used for identifying and comparing problem-solving skills and strategies used in solving problems.

How Can I Use These Materials When I Can't Even Finish What's in the Regular Textbook?

This is a common concern. But PSM is not intended to be an "add-on" program. Instead, much of PSM can replace material in the textbook. Correlation charts can be made suggesting how PSM can be integrated into the course of study or with the adopted text. Also, certain textbook companies have correlated their tests with the PSM materials.

Can the Materials Be Duplicated?

The pupil lessons may be copied for students. Each pupil lesson may be used as an overhead projector transparency master or as a blackline duplicator master. Sometimes the teacher may want to project one problem at a time for pupils to focus their attentions on. Other times, the teacher might want to duplicate a lesson for individual or small group work. Permission to duplicate pupil lesson pages for classroom use is given by the publisher.

How Can a Teacher Tell Whether Pupils Are Developing and Extending Their Problem-Solving Abilities?

Presently, reliable paper and pencil tests for measuring problem-solving abilities are not available. Teachers, however, can detect problem-solving growth by observing such pupil behaviors as

- identifying the problem-solving skills being used.

- giving accounts of successful strategies used in working on problems.
- insisting on understanding the topics being studied.
- persisting while solving difficult problems.
- working with others to solve problems.
- bringing in problems for class members and teachers to solve.
- inventing new problems by changing problems previously solved.

What Evidence Is There of the Effectiveness of PSM?

Although no carefully controlled longitudinal study has been made, evaluation studies do indicate that pupils, teachers, and parents like the materials. Scores on standardized mathematics achievement tests show that pupils are registering greater gains than expected on all parts of the test, including computation. Significant gains were made on special problem-solving skills tests which were given at the beginning and end of a school year.

Also, when selected materials were used exclusively over a period of several weeks with 6th-grade classes, significant gains were made on the word-problem portion of the standardized test. In general, the greater gains occurred in those classrooms where the materials were used as specified in the teacher commentaries and in-service materials.

Teachers have indicated that problem-solving skills such as *look for a pattern, eliminate possibilities*, and *guess and check* do carry over to other subjects such as Social Studies, Language Arts, and Science. Also, the materials seem to be working with many pupils who have not been especially successful in mathematics. And finally, many teachers report that PSM has caused them to make changes in their teaching style.

Why Is It Best to Have Whole-Staff Commitment?

Improving pupils' abilities to solve problems is not a short-range goal. In general, efforts must be made over a long period of time if permanent changes are to result. Ideally, then, the teaching staff for at least three successive grade levels should commit themselves to using PSM with their pupils. Also, if others are involved, this will allow for opportunities to plan together and to share experiences.

How Much In-Service Is Needed?

A teacher who understands the meaning of problem solving and is comfortable with the different
styles of teaching it requires could get by with
self in-service by carefully studying the section
and page commentaries in a grade-level book. The
different styles of teaching required include direct
instruction, guided discovery, laboratory work,
small group instruction, individual work, and nondirective instructions. The teacher would find the
audio tapes for each book and the *In-Service
Guide* a valuable resource and even a time saver.

If a school staff decides to emphasize problem
solving in all grade levels where PSM books are
available, in-service sessions should be led by
someone who has used the materials in the
intended way. For more information on this in-
service see the *In-Service Guide.*

What Materials Are Needed?

PROBLEM-SOLVING PROGRAM

REQUIRED MATERIALS	Grade 4	5	6	7	8	9
blank cards	X	X	X	X	X	X
bottle caps or markers	X			X		
calendar						X
calculators (optional for some activities)	X	X	X	X	X	X
cm squared paper, strips and singles						X
coins				X	X	
colored construction paper (circle fractions)	X	X				
cubes		X	X	X		X
cubes with red, yellow and green faces					X	
Cuisenaire rods (orange and white) or strips of paper		X				
dice (blank wooden or foam, for special dice)	X					
dice, regular (average 2 per student)	X	X	X	X	X	
geoboards, rubber bands, and record paper	X		X			
graph paper or cm squared paper			X	X		X
grid paper (1")			X			
metric rulers				X	X	X
phone books, newspapers, magazines		X		X		
protractors and compasses					X	
scissors	X	X	X	X		
spinners (2 teacher-made)			X			
tangrams	X					
tape measures				X		
thumbtacks (10 per pair of students)		X				
tile	X		X			
tongue depressors	X					
uncooked spaghetti or paper strips			X			

PSM   Rev.   1982

RECOMMENDED MATERIALS	Grade 4	5	6	7	8	9
adding machine tape				X		
centimetre rulers			X	X		
colored pens, pencils, or crayons		X				
coins, toy or real	X					
coins (two and one-half)						X
cubes					X	
demonstration ruler for overhead		X				
dominoes					X	
geoboard, transparent (for overhead)	X		X			
money - 20 $1.00 bills per student			X			
moveable markers		X	X		X	
octahedral die for extension activity				X		
overhead projector	X	X	X	X	X	X
place value frame and markers			X			
straws, uncooked spaghetti, or toothpicks		X				
transparent circle fractions for overhead	X	X				

PSM   Rev. 1982

Where Can I Find Other Problem Solving Materials?

RESOURCE BIBLIOGRAPHY

The number in parentheses refers to the list of publishers on the next page.

For students and teachers:

AFTER MATH, BOOKS I—IV by Dale Seymour, et al.
Puzzles to solve -- some of them non-mathematical.   (1)

AHA, INSIGHT by Martin Gardner
Puzzles to solve -- many of them non-mathematical.   (3)

THE BOOK OF THINK by Marilyn Burns
Situations leading to a problem-solving investigation.   (1)

CALCULATOR ACTIVITIES FOR THE CLASSROOM, BOOKS 1 & 2 by George Immerzeel and
Earl Ockenga
Calculator activities using problem solving.   (1)

GEOMETRY AND VISUALIZATION by Mathematics Resource Project
Resource materials for geometry.   (1)

GOOD TIMES MATH EVENT BOOK by Marilyn Burns
Situations leading to a problem-solving investigation.   (1)

FAVORITE PROBLEMS by Dale Seymour
Problem solving challenges for grades 5-7.   (3)

FUNTASTIC CALCULATOR MATH by Edward Beardslee
Calculator activities using problem solving.   (4)

I HATE MATHEMATICS! BOOK by Marilyn Burns
Situations leading to a problem solving investigation.   (3)

MATHEMATICS IN SCIENCE AND SOCIETY by Mathematics Resource Project
Resource activities in the fields of astronomy, biology, environment,
music, physics, and sports.   (1)

MIND BENDERS by Anita Harnadek
Logic problems to develop deductive thinking skills. Books A-1, A-2, A-3,
and A-4 are easy. Books B-1, B-2, B-3, and B-4 are of medium difficulty.
Books C-1, C-2, and C-3 are difficult.   (6)

NUMBER NUTZ (Books A, B, C, D) by Arthur Wiebe
Drill and practice activities at the problem solving level.   (2)

NUMBER SENSE AND ARITHMETIC SKILLS by Mathematics Resource Project
Resource materials for place value, whole numbers, fractions, and decimals.   (1)

The Oregon Mathematics Teacher (magazine)
Situations leading to a problem solving investigation.   (8)

PROBLEM OF THE WEEK by Lyle Fisher and William Medigovich
    Problem solving challenges for grades 7-12.   (3)

RATIO, PROPORTION AND SCALING by Mathematics Resource Project
    Resource materials for ratio, proportion, percent, and scale drawings.   (1)

STATISTICS AND INFORMATION ORGANIZATION by Mathematics Resource Project
    Resource materials for statistics and probability.   (1)

SUPER PROBLEMS by Lyle Fisher
    Problem solving challenges for grades 7-9.   (3)

For teachers only:

DIDACTICS AND MATHEMATICS by Mathematics Resource Project   (1)

HOW TO SOLVE IT by George Polya   (3)

MATH IN OREGON SCHOOLS by the Oregon Department of Education   (9)

PROBLEM SOLVING: A BASIC MATHEMATICS GOAL by the Ohio Department of Education   (3)

PROBLEM SOLVING: A HANDBOOK FOR TEACHERS by Stephen Krulik and Jesse Rudnik   (1)

PROBLEM SOLVING IN SCHOOL MATHEMATICS by NCTM   (7)

Publisher's List

1. Creative Publications, 3977 E Bayshore Rd, PO Box 10328, Palo Alto, CA 94303

2. Creative Teaching Associates, PO Box 7714, Fresno, CA 93727

3. Dale Seymour Publications, PO Box 10888, Palo Alto, CA 94303

4. Enrich, Inc., 760 Kifer Rd, Sunnyvale, CA 94086

5. W. H. Freeman and Co., 660 Market St, San Francisco, CA 94104

6. Midwest Publications, PO Box 448, Pacific Grove, CA 93950

7. National Council of Teachers of Mathematics, 1906 Association Dr, Reston, VA 22091

8. Oregon Council of Teachers of Mathematics, Clackamas High School, 13801 SE Webster St, Milwaukie, OR 97222

9. Oregon Department of Education, 700 Pringle Parkway SE, Salem, OR 97310

Grade 5

I. GETTING STARTED

I. GETTING STARTED

Teachers usually are successful at teaching skills in mathematics. Besides computation skills, they emphasize skills in following directions, listening, detecting errors, explaining, recording, comparing, measuring, sharing, ... They (You!) can also teach problem-solving skills. This section is designed to help teachers teach and students learn specific problem-solving skills.

I'm good at teaching skills — bet I could teach my pupils these problem-solving skills!

Some Problem-Solving Skills

Five common but powerful problem-solving skills are introduced in this section. They are:

. guess and check
. look for a pattern
. make a systematic list
. make and use a drawing or model
. eliminate possibilities

Students might use other skills to solve the problems. They can be praised for their insight but it is usually a good idea to limit the list of skills taught during the first few lessons. More problem-solving skills will occur in the other sections.

An Important DON'T

When you read the episodes that follow in this Getting Started section notice how the lessons are very teacher directed. The main purpose is to teach the problem-solving skills. Teachers should stress the skills verbally and write them on the board. Don't just ditto these activities and hand them out to be worked. Teacher direction through questions, summaries, praise, etc. is most important for teaching the problem-solving skills in this section. We want students to focus on specific skills which will be used often in all the sections. Later, in the Challenge Problems section, students will be working more independently.

Using the Activities

If you heed the important <u>Don't</u> on the previous page, you are on your way to success! The problems here should fit right in with your required course of study as they use whole number skills, elementary geometry and money concepts. In most cases, students will have the prerequisites for the problems in this section although you might want to check over each problem to be sure.

No special materials are required although markers, coins and boxes are helpful for some of the problems. The large type used for the problems makes them easier to read if they are shown on an overhead screen. In most cases students can easily copy the problem from the overhead. At other times you might copy the problem onto the chalkboard.

When and How Many

The <u>Getting Started</u> section should be used at the beginning of the year as it builds background in problem-solving skills for the other sections. As the format indicates <u>only one problem per day</u> should be used. Each should take less than twelve minutes of classtime if the direct mode of instruction is used. The remainder of the period is used for a lesson from the textbook or perhaps an activity from the <u>Drill and Practice</u> section of these materials.

> REMEMBER: One Problem Per Day when you are using this <u>Getting Started</u> section.

Guess And Check

The episode that follows shows how one teacher teaches the skill of guess and check. Notice that she very closely directs the instruction and constantly uses the terminology.

It is near the beginning of the year and Ms. Jones is about to start a math lesson. After getting the attention of the class, she begins.

Ms. J: I'm trying to figure out a number. Multiply the number by two.
Then add 11. The answer is 39. I wonder if the number is 10.
Is it?   (Waits for hands.)   Tim?

Tim: Nope!

Ms. J: How do you know?

Tim: 'Cause two times 10 is 20, add 11 gives 31.

Ms. J: By checking my guess you found out it was off. Well, is the
number 20?   Mary?

Mary: (Thinking out loud.)   Two times 20 is 40, add 11 is 51. No--too big.

Ms. J: Twenty is too big?   What can you say about 10?

Tim: It was too small.

Ms. J: Guessing and checking helped you decide 10 is too small and 20
is too big. Let's refine the guesses. Refine means to make
better guesses. Can you make a better guess?   (Some puzzled,
some thinking, some hands,)   Larry?

Larry: Try 15, it's in the middle.

Sue: (Computing on her paper.)   No, 15 is too big, it gives 41.

Ms. J: Does anyone think they know the answer? Scott?

Scott: 13

Ms. J: Why did you pick 13?

Scott: 39 is 2 less than 41 and 13 is 2 less than 15--Let's see--
2 times 13 is 26, add 11 gives 37. Oh! It's got to be 14!

Ms. J: Scott, you made a close guess, checked it and refined it to
get the right answer. Did you know that guess, check and
refine is a good way to solve problems? We're going to use
it a lot this year. I'm putting it up on the wall so we'll
all remember how important it is!

Let's try another problem.....

GUESS AND CHECK

WEEK 1 - DAY 1

a. I am thinking of a number. Multiply the number by 2.
    Then add 11. The answer is 39. What is the number?

b. I am thinking of a number. Subtract 5. Then add 18.
    The answer is 22. What is the number?

c. I am thinking of a number. Add the number to itself.
    Then subtract 13. The answer is 37. What is the number?

****

WEEK 1 - DAY 2

Find a path from start to finish with a sum of 74.
You may only go through the open gates.

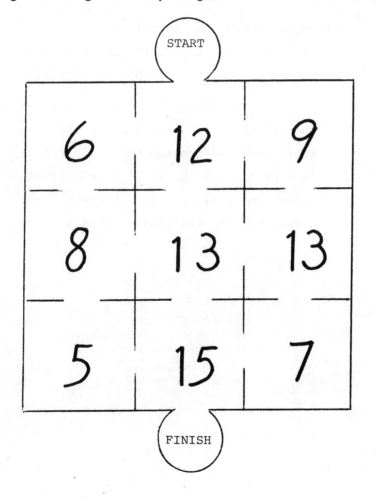

<u>Guess</u> <u>And</u> <u>Check</u>

Day 1. Answers: a. 14   b. 9   c. 25

Comments and suggestions:

. An outline of how to introduce these problems is given in Ms. Jones lesson on page 3. Remember, these are meant to be <u>teacher</u> <u>directed</u>.

. You can include more multiplication and division by using the problems below.

d. I am thinking of a number. Multiply the number by itself. Then subtract 13. The answer is 131. What is the number?   (12)

e. I am thinking of a number. Subtract 5. Then divide by 3. The answer is 22. What is the number?   (71)

Day 2. Answer: 12 → 6 → 8 → 13 → 13 → 7 → 15

Comments and suggestions:

. Begin by guessing a path you know is not correct. Have pupils check your path by mentally adding the values.

. If pupils do not realize that every path must go through the 12 and 15 squares, discuss with them how this knowledge would make the problem easier to refine.   12 + 15 = 27 and 74 - 27 = 47 so the problem is reduced to finding squares whose sum is 47. (Later this skill will be called "break a problem into manageable parts.)

Guess and Check (cont.)

WEEK 1 - DAY 3

For each problem, put the same number in the matching shapes to make true statements. Part _a_ is done for you.

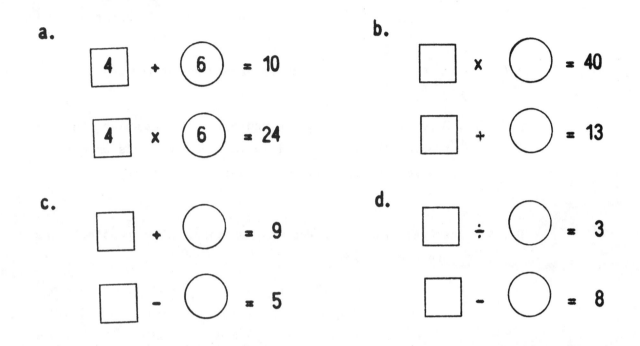

a.

$$\boxed{4} + \bigcirc{6} = 10$$

$$\boxed{4} \times \bigcirc{6} = 24$$

b.

$$\square \times \bigcirc = 40$$

$$\square + \bigcirc = 13$$

c.

$$\square + \bigcirc = 9$$

$$\square - \bigcirc = 5$$

d.

$$\square \div \bigcirc = 3$$

$$\square - \bigcirc = 8$$

****

WEEK 1 - DAY 4

Use any of the numbers 1, 2, 3, 4, 5, 6, 7, 8, 9 in the circles. Don't use the same number twice in any one problem.

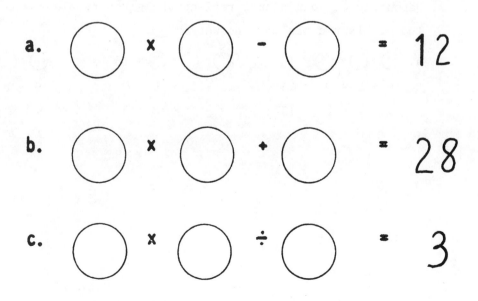

a. $\bigcirc \times \bigcirc - \bigcirc = 12$

b. $\bigcirc \times \bigcirc + \bigcirc = 28$

c. $\bigcirc \times \bigcirc \div \bigcirc = 3$

Day 3. Answers: a. 4, 6   b. 5, 8   c. 7, 2   d. 12, 4

Comments and suggestions:

. Pupils sometimes misunderstand this type of problem and
  use two numbers in the first part and two different
  numbers in the second part. Verbalizing the problem
  may help. For example, in part (a), this question could
  be asked,

    . Guess what two numbers have a sum of 10 and a
      product of 24.

. Depending on your class you may have to replace those
  problems containing multiplication and division with
  problems using only addition and subtraction.

Day 4. Answers: Each part has many answers. Some possibilities
        are given below.

a. $2 \times 8 - 4 = 12$ b. $3 \times 9 + 1 = 28$ c. $4 \times 6 \div 8 = 3$
    $3 \times 7 - 9 = 12$ $4 \times 5 + 8 = 28$ $2 \times 9 \div 6 = 3$
    $4 \times 5 - 8 = 12$ $3 \times 8 + 4 = 28$ $2 \times 6 \div 4 = 3$
    $2 \times 9 - 6 = 12$

Comments and suggestions:

. Ask pupils to give examples of guesses they tried which did
  not work. Pupils need to recognize that some problems can
  have many correct answers.

. Depending on your class, replace the given problems with
  these. Part c has many answers.

a. $\bigcirc + \bigcirc - \bigcirc = 15$ b. $\bigcirc - \bigcirc + \bigcirc = 16$ c. $\bigcirc - \bigcirc - \bigcirc = 3$
    $9 + 8 - 2 = 15$ $8 - 1 + 9$ $9 - 1 - 5$
    $9 + 7 - 1 = 15$ $9 - 1 + 8$ $9 - 2 - 4$
                                                    $9 - 4 - 2$

Guess and Check (cont.)

WEEK 1 - DAY 5

The numbers in the big circles
are found by adding the two
numbers in the small circles.
The example shows 14 = 6 + 8.
   16 = 10 + 6, and 18 = 8 + 10.

Example:

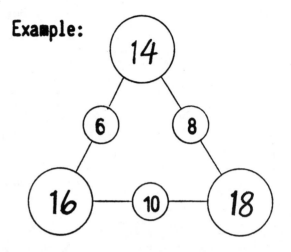

Find the numbers for the small circles in these two problems.

a.

b.

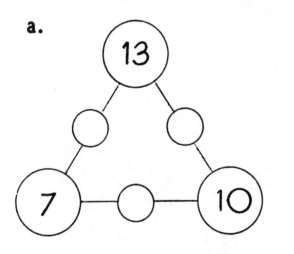

<u>Guess</u> <u>And</u> <u>Check</u>

Day 5. Answers: a. 5, 8, 2   b. 9, 14, 3

   Comments and suggestions:

   . Begin with a guess of three numbers such that no pair
     adds to any of the three given numbers. You might
     guess 2, 6 and 9.

   . The above guess helps pupils see that two of the
     numbers must add to one of the given numbers.
     Encourage pupils to guess possibilities that add
     to the smallest given number. For example, in (a)
     pupils might try 6 and 1, 5 and 2, or 4 and 3.

Look For A Pattern

By now your students are familiar with the skill guess and check.
One teacher introduced the next problem-solving skill, look for a
pattern, in this way.

Mr. Todd: Who remembers what method we used to solve problems last week?

Teri: We guessed.

Mr. Todd: Is that all?

Sam: Guess and check. It's up on the poster! (giggles)

Mr. Todd: That's right and we're going to add another problem-solving
skill to the poster today. (Writes it up.) What does it say?

Class: Look for a pattern.

Mr. Todd: Look for a pattern--this week we are going to practice looking
for patterns. Here's our problem. (Shows 2, 4, 6, ___, ___, ___
on overhead.) We want to fill in the next three blanks. Can
you see a pattern in the numbers?

Ellen: They're all even.

Sid: They go up by 2.

Mr. T: What goes in the blanks?

Several: 8, 10, 12

Mr. T: Let's try the next one. (Shows 1, 4, 7, 10, ___, ___, ___)
Who can tell me the numbers and the pattern they used?

Mary: 13, 16, 19. They go up by 3.

Mr. T: Good. Sue, you are frowning, what's wrong?

Sue: I saw odd, even, odd, even, but it didn't help to get the numbers.

Mr. T: Well, it is a good pattern though. That's pretty neat to remember
about odd and even. Sometimes one pattern isn't enough to solve
a problem and we have to look for another one. Let's see if we
can solve the rest ...

LOOK FOR A PATTERN

WEEK 2 - DAY 1

Write the next three numbers in each sequence.

a.   2, 4, 6, ___, ___, ___

b.   1, 4, 7, 10, ___, ___, ___

c.   1, 2, 4, 8, 16, ___, ___, ___

d.   1, 2, 4, 7, 11, 16, ___, ___, ___

****

WEEK 2 - DAY 2

Find a rule that gives the third number from the first two numbers. Fill in the blanks.

a.			b.			c.		
8,	3,	11	5,	2,	10	10,	4,	6
9,	5,	14	7,	5,	35	15,	7,	8
4,	8,	___	6,	4,	___	19,	9,	___
6,	___,	20	9,	___,	45	7,	___,	1
___,	7,	19	___,	10,	30	___,	11,	12

PSM 81

Look For A Pattern

Day 1. Answers: a. 8, 10, 12   b. 13, 16, 19

c. 32, 64, 128   d. 22, 29, 37

Comments and suggestions:

. An outline of how to introduce these problems is given
in Mr. Todd's lesson on page 13.

. Pupils usually like to look for patterns like these.

. Occasionally they will complete patterns with other
answers. If a good argument is provided to support
these answers, credit should be given.

Day 2. Answers:

a.	add	b.	multiply	c.	subtract
	4, 8, <u>12</u>		6, 4, <u>24</u>		19, 9, <u>10</u>
	6, <u>14</u>, 20		9, <u>5</u>, 45		7, <u>6</u>, 1
	<u>12</u>, 7, 19		<u>3</u>, 10, 30		<u>23</u>, 11, 12

Comments and suggestions:

. Occasionally pupils will complete patterns with other
answers. For example, problem (b) looks as if the
answers must end in 0 or 5. Or in problem (c), pupils
might say the second number is smaller than either the
first or third number. If a good argument is provided
to support the pattern, credit should be given.

Look For A Pattern (cont.)

WEEK 2 - DAY 3

$$1 \times 8 + 1 = \underline{\hspace{2cm}}$$
$$12 \times 8 + 2 = \underline{\hspace{2cm}}$$
$$123 \times 8 + 3 = \underline{\hspace{2cm}}$$
$$1{,}234 \times 8 + 4 = \underline{\hspace{2cm}}$$

a.   Predict the answer for $123{,}456 \times 8 + 6$.   \underline{\hspace{3cm}}

b.   Check your prediction.

c.   Predict and check $123{,}456{,}789 \times 8 + 9$ \underline{\hspace{3cm}}

****

WEEK 2 - DAY 4

These are stair-step numbers: 1, 3, 6, ...

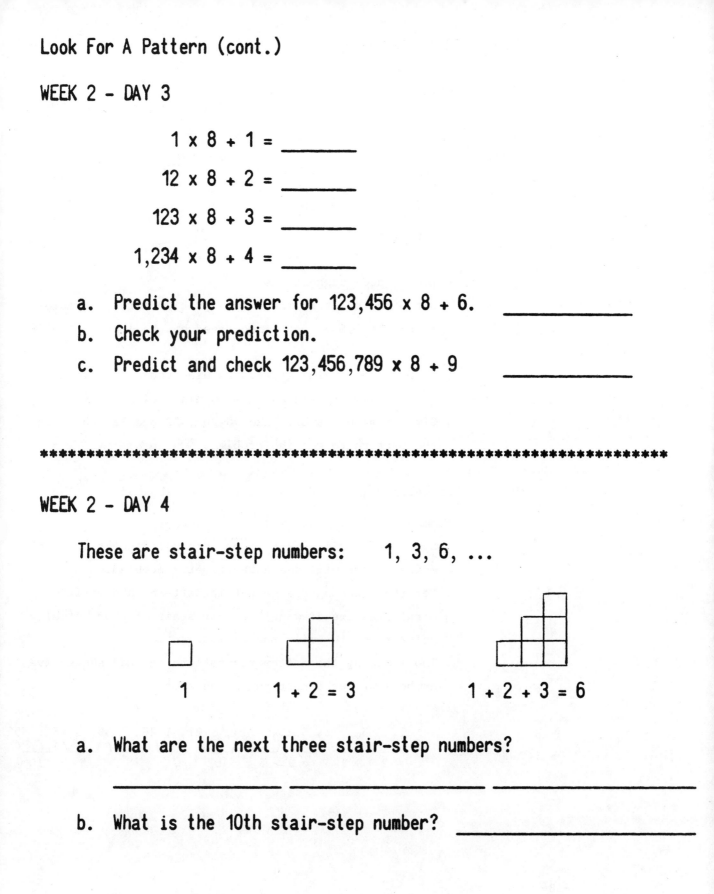

1   1 + 2 = 3   1 + 2 + 3 = 6

a.   What are the next three stair-step numbers?

\underline{\hspace{4cm}} \underline{\hspace{4cm}} \underline{\hspace{4cm}}

b.   What is the 10th stair-step number?   \underline{\hspace{5cm}}

<u>Look</u> <u>For</u> <u>A</u> <u>Pattern</u>

Day 3.   Answers:

$$1 \times 8 + 1 = \underline{9}$$
$$12 \times 8 + 2 = \underline{98}$$
$$123 \times 8 + 3 = \underline{987}$$
$$1{,}234 \times 8 + 4 = \underline{9{,}876}$$

a. 987,654

b. Prediction checks

c. 987,654,321

Comments and suggestions:

. Use of calculators makes this problem more enjoyable
  to pupils. Some get discouraged by the long multipli-
  cations.

. Pupils should give answers as nine thousand, eight
  hundred seventy-six and not as nine, eight, seven, six.

. Give them an answer, like 98,765,432 and let them use
  the pattern to get the problem.

Day 4.   Answers: a. 10, 15, 21   b. 55

Comments and suggestions:

. Some pupils will see this pattern as a geometric problem
  while others will see an arithmetic problem.

. For those seeing the geometric pattern, appropriate
  strategies are drawing the next stair-step or actually
  using tiles to build the stairs.

. Those seeing the arithmetic pattern can add successive
  numbers or see a pattern in the sums.

```
1,   3,   6,   10,   15,   21,   28,   36,   45,   55
  \_/  \_/  \_/   \_/    \_/    \_/    \_/    \_/    \_/
   2    3    4     5      6      7      8      9     10
```

Look For A Pattern (cont.)

WEEK 2 – DAY 5

Three figures are given.   Sketch the next two figures in each sequence.

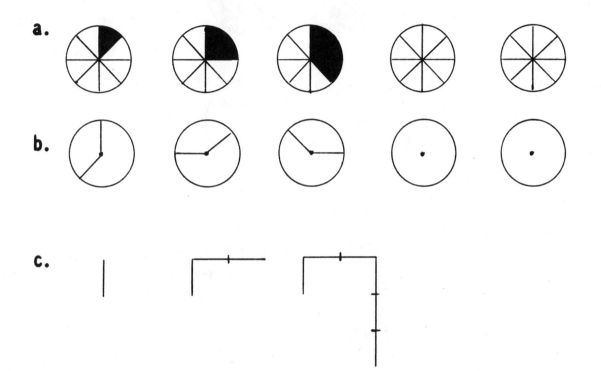

a.

b.

c.

© PSM 81

Day 5. Answers:

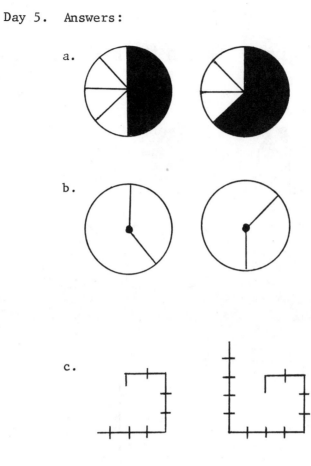

a.

b.

c.

Comments and suggestions:

. Pattern (b) may be related to the hands of a clock.

. Pattern (c) could have several different answers depending
  upon pupil perceptions. For example, some may extend
  the fourth segment to the right instead of spiraling
  around as the answer shows.

Make A Systematic List

Mr. Benton: What is our new problem-solving skill? (Students read from poster.) Great--here's a problem to try it on. (Shows problem on overhead and makes sure pupils understand it.)

> Carla and Nick were playing a factors game. Carla would name a factor of 24 and Nick had to give the other factor. Show the ways Carla and Nick could play the game.

Mr. B: What are some of the ways Carla and Nick can play the game? Scott? 3 and 8. Good. Sue? 4 and 6. Joe? 2 and 12. Any more? (Silence.) Do we have them all? Would it help to make a systematic list? Which pair shall we write first?

Carla	Nick
2	12
3	8
4	6

Ted: 2 and 12, 'cause 2 is smallest, then 3 and 8, then 4 and 6.

Mr. B: You know, I can't tell which number was chosen by Carla. Maybe we'd better label our list. Now, did we miss any?

Mary: Well, we could switch them--you know, Carla 6 and Nick 4.

Mr. B: That gives us six solutions. Could Carla's numbers get smaller?

James: No--Oh yes! 1 and 24 or 24 and 1.

Mr. B: Great. Do you see how making a systematic list or table can help solve a problem? Let's try another...

MAKE A SYSTEMATIC LIST

WEEK 3 - DAY 1

Carla and Nick were playing a factors game. Carla would name a factor of 24 and Nick had to give the other factor. Show the ways Carla and Nick could play the game.

Carla's factor								
Nick's factor								

****

WEEK 3 - DAY 2

You have the three letters A, E, and T. List all the 3-letter combinations you can make using each letter once. How many of the combinations are actual words?

****

WEEK 3 - DAY 3

At camp these are the choices for supper.

Meat	Potatoes	Vegetable
steak	mashed	corn
trout	baked	green beans
	french fries	

List the 12 different suppers a camper could choose.

Make A Systematic List

Day 1. Answers:

Carla's factor	1	2	3	4	6	8	12	24
Nick's factor	24	12	8	6	4	3	2	1

Comments and suggestions:

. An outline on how to introduce these lessons is given in Mr. Benton's lesson on page 21.

. Pupils will usually begin by listing the factors at random. Systematically listing the factors in the table usually convinces pupils that all pairs have been found.

. Pupils should see that as Carla's factors increase, Nick's factors decrease.

Day 2. Answers: AET, ATE, EAT, ETA, TAE, TEA. Only TAE is not listed in Webster's New Collegiate Dictionary. AET is an abbreviation meaning aged. ETA is the seventh letter of the Greek alphabet.

Comments and suggestions:

. Slips of paper with the letters written on them can be moved around to find the different letter combinations.

. Pupils may discover three ways to keep the first letter constant and switch the other two, e.g. A then E T or T E .

Day 3. Answers: smc   sbc   sfc   tmc   tbc   tfc
                   smg   sbg   sfg   tmg   tbg   tfg

Comments and suggestions:

. Encourage the use of some kind of "abbreviation" for making the list.

. If pictures are available, actually showing the different meals would make an effective bulletin board.

. A tree diagram (partially displayed below) makes another visual display of the problem.

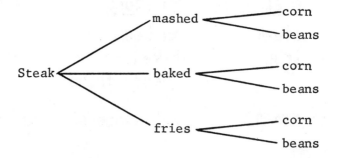

Make A Systematic List (cont.)

WEEK 3 - DAY 4

Pat was playing "Guess The Number" and gave these clues:

a. The number has 3 digits.
b. The digit in the 100's place is greater than 7.
c. The digit in the 10's place is less than 2.
d. The number is even.

What are the possibilities for the number?

Pat gave this additional clue:

e. The sum of the 3 digits is 12.

Now what are the possibilities for the number?

****

WEEK 3 - DAY 5

The coins shown above are the only ones you have.
What amounts can you make if you use 1, 2, 3, or 4 of
the coins?

Make A Systematic List

Day 4. Answers:

1st question:				2nd question:
800	810	900	910	
802	812	902	912	912 and 804
804	814	904	914	
806	816	906	916	
808	818	908	918	

Comments and suggestions:

. Each clue limits the numbers to include in the systematic list.

. Have pupils create a third clue that limits the answer to just one possibility. An example would be: All three digits are even which makes 804 the only answer.

Day 5. Answers:

	1¢	5¢	10¢	25¢	Total
a.	✓				1¢
		✓			5¢
			✓		10¢
				✓	25¢
b.	✓	✓			6¢
	✓		✓		11¢
	✓			✓	26¢
		✓	✓		15¢
		✓		✓	30¢
			✓	✓	35¢
c.	✓	✓	✓		16¢
	✓	✓		✓	31¢
	✓		✓	✓	36¢
		✓	✓	✓	40¢
d.	✓	✓	✓	✓	41¢

Comments and suggestions:

. Fifteen different amounts can be shown. The way the problem is stated excludes none of the coins or 0¢., which would make a sixteenth way.

. Emphasize the pattern used to make the systematic list, i.e., all combinations of just one coin (a), all combinations of two coins (b), all combinations of three coins (c), then the combination using all four coins (d).

Make And Use A Drawing Or Model

Usually the problem-solving skills in this section should be introduced directly, but sometimes they can be brought out by taking advantage of pupil efforts. This problem was given in the fourth week of problem-solving in Ms. Thompson's class:

A fireman stood on the middle step of a ladder. As the smoke got less, he climbed up three steps. The fire got worse so he had to climb down five steps. Then he climbed up the last six steps and was at the top of the ladder. How many steps were in the ladder?

Pupils were allowed to think about the problem for a few minutes. Soon some thought they had solved it, others weren't sure and some had given up.

Ms. T: Nora, you wrote down some things to use on the problem. Will you share what you did?

Nora: Well, I guessed 7 steps and counted up and down on the steps but it didn't work.

Ms. T: You used a <u>guess</u> <u>and</u> <u>check</u> approach. That's good but I also see you drew a ladder on your paper.

Nora: I had to so I could count the steps and keep track.

Ms. T: I saw several of you making a drawing on your paper. Do you know what our next problem-solving skill is going to be? <u>Make</u> <u>and</u> <u>use</u> <u>a</u> <u>drawing</u> <u>or</u> <u>model</u>. That sounds long and complicated but some of you are already doing it. Let's look at some of the drawings you tried. Tom, will you describe what you did and I'll try to copy it on the board.

Tom: I drew 5 steps and marked the middle. Then I counted up 3, but I had to add a step. I counted down 5 and up 6 and added another step. That gave 7 so I thought it was 7 steps.

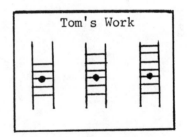

Tom's Work

Ms. T: Do you still think it's 7?

Tom: When Nora said it didn't work, I saw that my middle mark was off.

Ms. T: Nora used a drawing to check her guess; Tom used a drawing to see why an answer wasn't right. Making a drawing is pretty helpful. Has anyone found the answer? 9? Let's <u>make</u> <u>a</u> <u>drawing</u> and check it ...

WEEK 4 - DAY 1

A fireman stood on the middle step of a ladder.
As the smoke got less, he climbed up three steps.
The fire got worse so he had to climb down five steps.
Then he climbed up the last six steps and was at the top
of the ladder.

How many steps were in the ladder?

****

WEEK 4 - DAY 2

One way to wrap a package is shown. How much ribbon is needed
to wrap the package if it is 60 cm long, 40 cm wide, and 20 cm
high?

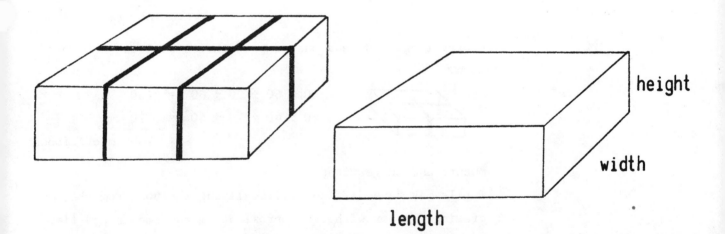

Show a different way the same package could be wrapped to use
280 cm of ribbon.

PSM 81

Make And Use A Drawing Or Model

Day 1. Answer: 9 steps

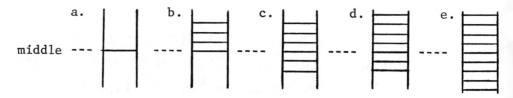

Comments and suggestions:

. An outline on how to introduce these lessons is given in
  Ms. Thompson's lesson on page 27.

. A drawing helps solve this problem. The sequence of drawings
  illustrate the clues.

. Part (e) shows four steps above the middle step so there are
  four steps below the middle step.

. Be sure pupils do not count the step the fireman is standing
  on as a step when he moves up.

Day 2. Answers:

One wrap lengthwise   =   20 + 60 + 20 + 60   =   160
Two wraps widthwise   =   20 + 40 + 20 + 40   =   120
                          20 + 40 + 20 + 40   =   120
                                                  ─────
                                                  400 cm of ribbon

One wrap lengthwise and one wrap widthwise will use 280 cm of
ribbon.

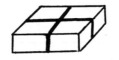

20 + 60 + 20 + 60   =   160
20 + 40 + 20 + 40   =   120
                        ─────
                        280 cm of ribbon

Comments and suggestions:

. Pupils may have difficulty visualizing the box from the
  drawing. A box with the approximate measurements and tied
  with ribbon may help in visualizing.

Make And Use A Drawing Or Model (cont.)

WEEK 4 - DAY 3

From a large sheet of postage stamps, three attached stamps can be torn off in six different ways. Make drawings to show the six ways.

****

WEEK 4 - DAY 4

In a horse race:

a. Crazyhorse finished 1 length ahead of Appleater.
b. Appleater was not the last place horse.
c. BobbiSue finished 7 lengths ahead of Electricity.
d. Crazyhorse finished 7 lengths behind Dobbin.
e. Electricity finished 3 lengths behind Crazyhorse.

What was the finishing position of each horse?

****

WEEK 4 - DAY 5

Two hunters came to a river. They met three youngsters who had a small boat. The boat could hold two youngsters or one adult. How did all five persons get across the river? How many trips will be necessary?

PSM 81

Day 3. Answers: Six distinct ways are possible.

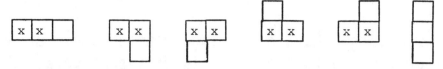

Comments and suggestions:

. Grid paper, Christmas seals or Green Stamps are choices for the model to use. If grid paper is used, suggest to pupils that they draw faces on each stamp.

. Note the pattern shown in the first five drawings. Two stamps are noted, then the third stamp is located (1) at the right end, (2) lower right, (3) lower left, (4) upper left, and (5) upper right. The left end is not used as this way would not be different than the right end model.

Day 4. Answer: 1st - Dobbin, 2nd - BobbiSue, 3rd - Crazyhorse
4th - Appleater, 5th - Electricity

Comments and suggestions:

. A line divided into sections makes a convenient drawing for this problem.

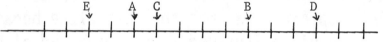

. Perhaps actual pictures of horses could be placed along a line to simulate the finish of the race. Pupils might focus on the horse mentioned most often. What information is known about this horse?

Day 5. Answer: Eleven trips are needed.

Start	Trip	HHYYY		
	1	HHY	YY --→	YY
	2	HHYY	←-- Y -	Y
	3	HH	-- YY →	YYY
	4	HHY	←- Y --	YY
	5	H Y	--- H →	YY H
	6	H YY	←- Y -	Y H
	7	YY	--- H →	Y HH
	8	YYY	←-- Y -	HH
	9	Y	-- YY →	YY HH
	10	YY	←- Y -	Y HH
	11		-- YY →	YYYHH

Comments and suggestions:

. Pupils could use markers to move back and forth over an imaginary river.

. Or five pupils could actually walk through the problem.

. In either case, record keeping is important to avoid repeating false starts.

. Some of the trips shown could be done in a different order, but eleven trips are still necessary.

Eliminate Possibilities

Usually pupils try to solve a problem by looking directly for the answer, but sometimes it is more helpful to identify or list possible answers and then eliminate incorrect answers. This narrows the search for a correct answer and in some cases leaves only the correct answer.

Mr. Coal used this problem to introduce the skill eliminate possibilities:

Find four whole numbers to put in the squares. Each pair must have the sum given.

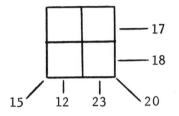

Mr. C: Here is a problem you could solve by guess and check, but I am going to have you solve it a little differently. Let's look at one of the squares. (Points to 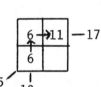) What is the greatest number we can have in this square?

Donna: 12, if we allow zero.

Mr. C: Zero is fine - it's a whole number. 12 is the greatest. What are all the possible numbers for this square?

Wayne: Any whole number from 0 to 12.

Mr. C: (Writes 0,1,2,3,4,5,6,7,8,9,10,11,12) Good. Let's see which of these numbers we can eliminate. Suppose we try 0 in the square; then we have this ⟶ and the diagonal only adds up to five. (Crosses off 0 and 12 from the list.)

Jane: With 1, the diagonal only adds up to 7. (Crosses off 1 and 11 from the list.) Let's skip up and try 6.

Mr. C: With 6 in the square we have 17 for the diagonal. (Crosses off 6.)

Pete: 5 gives 15 for the diagonal and, let's see, that's 13 in the last square--Yes, it checks.

Mr. C: Did it help to make a list of the possibilities and eliminate those that didn't work?

Toby: It helped me to see that I didn't have to try more than those numbers and as you cross some off you know you are getting closer to the solution.

Mr. C: Why didn't I pick one of the other squares?

Joyce: The list of numbers would have been larger.

Mr. C: Right. This week we are going to solve problems using the skill eliminate possibilities.

ELIMINATE POSSIBILITIES

WEEK 5 - DAY 1

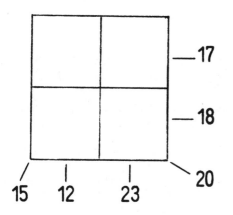

Find four whole numbers to put in the square. Each pair must have the sum given.

****

WEEK 5 - DAY 2

Jeff has less than 30 marbles.

When he puts them in piles of 3 he has no marbles left over.

When he puts them in piles of 2 he has 1 left.

When he puts them in piles of 5 he has 1 left.

How many does he have?

****

WEEK 5 - DAY 3

I am a number less than 100.

My units digit is a 4.

The sum of my digits is an odd number.

My tens digit is a multiple of 3.

Who am I?

-35-

Eliminate Possibilities

Day 1.   Answers:

7	10
5	13

Comments and suggestions:

. An outline on how to introduce these lessons is given in
  Mr. Coal's lesson on page 33.

. Be sure pupils see the sums of the two diagonals.

Day 2.   Answer: 21

Comments and suggestions:

. Using the first and second clues, the possibilities are
  0, 3, 6, 9, 12, 15, 18, 21, 24, 27.

. The third clue eliminates all the evens leaving 3, 9, 15,
  21 and 27.

. The fourth clue eliminates all others except for the
  answer 21.

Day 3.   Answers: 34 and 94.

Comments and suggestions:

. Possibilities using the first two clues are 4, 14, 24, 34,
  44, 54, 64, 74, 84 and 94.

. The third clue eliminates all except 14, 34, 54, 74 and 94.

. The fourth clue eliminates 14, 54 and 74. Only 34 and 94
  remain.

Eliminate Possibilities (cont.)

WEEK 5 – DAY 4

Use the digits 3, 4, 8, and 9. Make two addition problems so each sum is greater than 150.

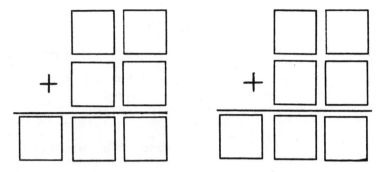

****

WEEK 5 – DAY 5

```
  A A A
  B B B
+ C C C
-------
  D D D
```

A, B, C, and D each stand for a different digit.

Which of these are not possibilities for DDD?

111   222   333   444   555   666

Eliminate Possibilities

Day 4. Answers:

```
    83        84
  + 94      + 93
  ----      ----
   177       177
```

Comments and suggestions:

. Eliminate 3 and 4 as possibilities for the tens digits
  because the sum will not be greater than 150.

. Pupils may give
```
   83
 + 94
```
and
```
   94
 + 83
```
as different.

Discuss with them why these are the same.

Day 5. Answers: 111 and 222

Comments and suggestions:

. To have D = 1 or D = 2 means A, B, and C can not be
  a 1 or a 2. So a possibility is 333 + 444 + 555.
  But this requires "carrying" and the sum becomes 332.
  Even using 000 does not allow 111 or 222 as possible sums.

. All others are possible using 000. For example,
  000 + 111 + 222 = 333; 000 + 111 + 333 = 444;
  000 + 222 + 333 = 555; and 111 + 222 + 333 = 666.

Grade 5

II.   WHOLE NUMBER DRILL AND PRACTICE

II. WHOLE NUMBER DRILL AND PRACTICE

Most fifth-grade classes are a collection of students with varying levels of skills. What can be done to provide additional practice and learning for all students? One solution is to offer whole number drill and practice through problem-solving activities. While Tim is remembering that 3 x 8 is 24, not 32, Hosea might be figuring out all the possible ways to complete

Topic for the week: Whole Number Multiplication

Oh no! I've forgotten all the facts.

I learned all that stuff last year!

_____ x _____ = 24. Another solution is to use games. If the games are combined with problem solving, many objectives are accomplished at the same time; motivation, computation practice, and emphasizing of problem-solving skills.

One marvelous discovery in using problem-solving skills with a mixed class is that some students who are unskilled in computation are really good at seeing patterns or figuring out ways to solve problems. Some who hate ordinary drill problems will fill a page with computation to try to solve a problem.

Using the Activities

The activities in this section can be incorporated with the regular teaching and review of whole number operations. Realizing that students have different skills, many activities have suggestions for simplifying or extending the problems. Many of the activities are meant to be teacher directed while others can be finished by students after a brief introduction.

DIGIT DRAW ACTIVITIES
(Ideas for Teachers)

Ten digit cards marked 0-9 can be used for a variety of activities.

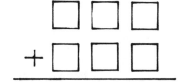

A. Have pupils make the diagram to the right.

B. Suggest a goal, such as, getting the largest possible sum.

C. Shuffle and draw digits, one at a time. Pupils must write each digit, as it is drawn, until all spaces are filled.

D. Compare and discuss results.

This activity is very adaptable. For example, in the 3-digit by 3-digit addition problem above, pupils may not get the largest sum the first time. The teacher can have the pupils use the same six digits to find the largest sum; then use the same six digits to find the smallest sum; and then use those digits to find the sum closest to, say, 700.

These activities can be used for drill and practice, concept development and/or diagnosis. Each is highly individual, as many pupils will have a unique problem. See the next page for other suggested formats. Further variations could include:

  . replacing a drawn digit so it can be used again

  . restricting certain numbers, e.g., no zero is used with division and fraction problems

  . allowing a special reject box giving pupils the chance to discard an unfavorable draw

  . adapting whole number activities to decimal activities by inserting a decimal point(s).

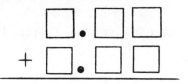

PSM 81

Digit Draw Activities

Mathematics teaching objectives:

. Develop place value concepts.

. Develop informal probability concepts.

. Practice computation skills.

Problem-solving skills pupils _might_ use:

. Break problem into manageable parts.

. Make decisions based upon data.

. Recognize limits and/or eliminate possibilities.

Materials needed:

. Digit cards 0-9

Comments and suggestions:

. This activity works well as an opener at the beginning of the class or as an ending for that last few minutes of a class when all other activities have been completed. See pages 43 and 45 for other suggestions and many variations.

. After some trials, pupils can be encouraged to share what strategies they use. "If a 9 or 8 is drawn first, put it in the 100's place. If a little digit comes first, put it in the 1's place. It's hard to decide what to do with a 5 or 6. Sometimes it's just luck!"

. The strategies pupils use to solve problems 1, 2 and 3 on the second page will very greatly. In problem 3 one student might write out lots of possibilities to see which one gives smaller answers. This student needs encouragement in organizing the trials and making conclusions from the trials that give larger answers. Another might try to solve the problem by analyzing the subtraction process. Others might take the problem to their parents or an older friend for some shared problem solving. Students can be complimented on their efforts even if they don't find the smallest difference. The smallest difference found could be posted and changed as someone finds a smaller difference.

Answers:

Answers will vary according to the digits drawn and the places where pupils put the digits.

Digit Draw Activities (cont.)

OTHER FORMATS

Place Value: □,□ □ □ largest number, smallest number, or
                       closest to 5000

Ordering: □ □ < □ □ < □ □

Addition: □ □   □ □ □   Subtraction: □ □ □   □ □ □
           □ □ +□ □ □   - □ □   -□ □ □
          +□ □

Multiplication: □ □ □   □ □ □ □   □ □   □ □ □
                  × □   × □   ×□ □   ×□ □

Division: □⟌□ □ □   □⟌□ □ □ □   □ □⟌□ □ □ □

Fractions: $\dfrac{\Box}{\Box} + \dfrac{\Box}{\Box}$ or any other operation

Two other activities, seemingly obvious, lead to an interesting third problem.

1. Using each digit once, make the largest possible 10-digit number.
2. Using each digit once, make the largest possible sum for a 5-digit by 5-digit addition problem.
3. Using each digit once, make the smallest possible difference for a 5-digit by 5-digit subtraction problem.

The answer to (1) is 9,876,543,210. The answer to (2) is 97,531 + 86,420 = 183,951. Many pupils will think the answer to (3) is 97,531 - 86,420 = 11,111. But a much smaller difference is possible--50,123 - 49,876 = 247.

PSM 81

H's ON THE HUNDREDS GRID

0	1	2	③	4	⑤	6	7	8	9
10	11	12	⑬	⑭	⑮	16	17	18	19
20	21	22	㉓	24	㉕	26	27	28	29
30	31	32	33	34	35	36	37	38	39
40	41	42	43	44	45	46	47	48	49
50	(51)	52	(53)	54	55	56	57	58	59
60	(61)	(62)	(63)	64	65	66	67	68	69
70	(71)	72	(73)	74	75	76	77	78	79
80	81	82	83	84	85	86	87	88	89
90	91	92	93	94	95	96	97	98	99

The top H is called the 14-H. The bottom H is called the 62-H.

1. Draw the 87-H.
2. Find the sum of the numbers in the
   a. 14-H b. 62-H c. 87-H
3. Find an H with a sum equal to 147.
4. How can you tell when the H-sum will be even or odd?
5. Write what you think the 20-H is.
6. Find the sum of the numbers in the
   a. 11-H b. 20-H c. 12-H
7. Discover a way to find the sum if you know the middle number of an H.
8. Discover a way to find the middle number of an H if you know the sum.

<u>H's On the Hundreds Grids</u>

Mathematics teaching objectives:

. Practice computation skills.

. Use a calculator (optional).

Problem-solving skills pupils <u>might</u> use:

. Look for patterns.

. Make predictions based upon data.

Materials needed:

. Calculator (optional)

Comments and suggestions:

. If your mathematics objective is <u>not</u> to practice basic skills, a
calculator can be used to do the computations for this investigation.

. The discovery in problem 6 can be made by examining "easier" cases;
for example, by studying the results of 11-H and 12-H. This hint
may be necessary to give to some pupils.

. An interesting extension involves making a "sideways H", e.g.

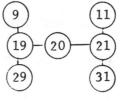

The same discoveries hold for these shapes.

Answers:

1. Pupil drawing.

2. a. 98   b. 434   c. 609

3. 21-H

4. An even middle number gives an even H-sum. An odd middle number gives
an odd H-sum.

5. Answers will vary. Most will say—

6. a. 77   b. 140   c. 84

7. Multiply the middle number by 7.

8. Divide the sum by 7.

CATCH 22

1. a. Choose two different digits. ____ ____
   b. Form two different 2-digit numbers. _____
   c. Add the two 2-digit numbers. _____
   d. Add the two digits you chose. _____
   e. Divide the answer to c by the answer to d. _____
   f. Repeat steps a-e with two different digits.
   g. What did you discover? _____

2. a. Choose three different digits. ____ ____ ____
   b. Form six different 2-digit numbers from the digits you chose.

      ____ ____ ____ ____ ____ ____
   c. Add the six numbers. _____
   d. Add the three digits you chose. _____
   e. Divide the answer to c by the answer to d. _____
   f. Repeat steps a-e with three different digits.
   g. What did you discover? _____

3. a. Suppose you chose four different digits and performed the same type of investigation. What answer do you think you would get? _____
   b. Do it now. ____ ____ ____ ____
   c. Form twelve different 2-digit numbers. ____ ____ ____

      ____ ____ ____ ____ ____ ____ ____ ____ ____
   d. Continue.

4. Predict the "magic" number for an investigation using eight different digits. _____

PSM 81

<u>Catch 22</u>

Mathematics teaching objectives:

. Practice computation skills.

Problem-solving skills pupils <u>might</u> use:

. Look for a pattern.

. Make a systematic list.

Materials needed:

. None

Comments and suggestions:

. Some pupils will need help making the systematic list so all the different two digit numbers are found in problems 2, 3, and 4.

. Pupils usually can find the pattern and predict the result in problem 4.

. Some of your better pupils can be challenged to prove the investigation using eight digits, as in number 4.

Answers:

1. g. The result is always 11

2. g. The result is always 22

3. g. The result is always 33

4. 77

For your benefit only, this is why the trick works for three digits. Using the three digits a, b, c

$$10a + b$$
$$10a + c$$
$$10b + a$$
$$10b + c$$
$$10c + a$$
$$\underline{10c + b}$$
$$22a + 22b + 22c$$

and

$$\frac{22a + 22b + 22c}{a + b + c} = \frac{22(a + b + c)}{a + b + c} = 22$$

SUBTRACTION INVESTIGATIONS

1. a. Make a two-digit number using consecutive digits. _____

   b. Reverse the number. _____

   c. Subtract the two numbers. _____

   d. Repeat steps a-c twice more. What did you discover?

   _____

2. a. Make a three-digit number using consecutive digits. _____

   b. Reverse the number. _____

   c. Subtract the two numbers. _____

   d. Repeat steps a-c twice more. What did you discover?

   _____

3. a. Make a four-digit number using consecutive digits. _____

   b. Reverse the number. _____

   c. Subtract the two numbers. _____

   d. Repeat steps a-c twice more. What did you discover?

   _____

4. a. Choose two different digits. _____ _____

   b. Make the largest and smallest numbers using the two digits.

   _____ _____

   c. Subtract the two numbers. _____

   d. Repeat steps b and c with the difference until something
      special happens. Describe what happened.

   _____

PSM 81

Subtraction Investigation

Mathematics teaching objectives:

    . Practice subtraction skills.

Problem-solving skills pupils <u>might</u> use:

    . Look for a pattern.

    . Make generalizations based on observed patterns.

Materials needed:

    . None

Comments and suggestions:

    . Pupils may need an example of each of the numbers asked for in problems 1 — 3, e.g. 32 or 543 or 9876.

    . The investigation in problems 1 — 3 involves just one subtraction. But in problems 4 — 6, several subtractions may be needed before "something special" happens. Be sure pupils understand that the digits in the difference are used to make the next largest and smallest numbers.

Answers:

1. The difference is always 9.

$$\begin{array}{r} 54 \\ -\ 45 \\ \hline 9 \end{array}$$

2. The difference is always 198.

$$\begin{array}{r} 654 \\ -\ 456 \\ \hline 198 \end{array}$$

3. The difference is always 3087.

$$\begin{array}{r} 9876 \\ -\ 6789 \\ \hline 3087 \end{array}$$

4. The difference will eventually become 9.

5. The difference will eventually become 495.

6. The difference will eventually become 6174. Using 3, 1, 8, and 7 the problem becomes

$$\begin{array}{r} 8731 \\ -\ 1378 \\ \hline 7353 \end{array} \rightarrow \begin{array}{r} 7533 \\ -\ 3357 \\ \hline 4176 \end{array} \rightarrow \begin{array}{r} 7641 \\ -\ 1467 \\ \hline 6174 \end{array} \rightarrow \begin{array}{r} 7641 \\ -\ 1467 \\ \hline 6174 \end{array}$$

Subtraction Investigations (cont.)

5. a. Choose three different digits. _____ _____ _____

    b. Make the largest and smallest numbers using the three digits.

       _____ _____

    c. Subtract the two numbers. _____

    d. Repeat steps <u>b</u> and <u>c</u> with the difference until something special happens. Describe what happened.

    _____

6. a. Choose four different digits. _____ _____ _____ _____

    b. Make the largest and smallest numbers using the four digits.

       _____ _____

    c. Subtract the two numbers. _____

    d. Repeat steps <u>b</u> and <u>c</u> with the difference until something special happens. Describe what happened.

    _____

PSM 81

MULTIPLICATION PATTERNS

1. Fill in the horizontal section. Look for a pattern.
2. Fill in the vertical section. Look for a pattern.
3. Fill in the diagonal section. Look for a pattern.
4. Fill in the zig-zag section. Look for patterns.

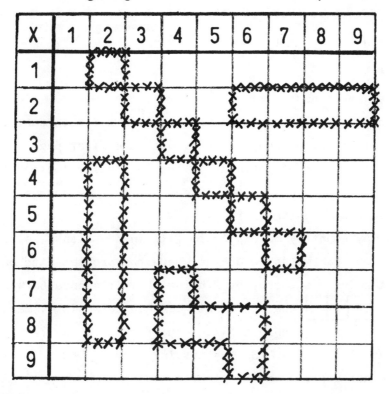

5. Now fill in the rest of the table. Use patterns if you can.

6. Find these sections. Shade them lightly as shown.

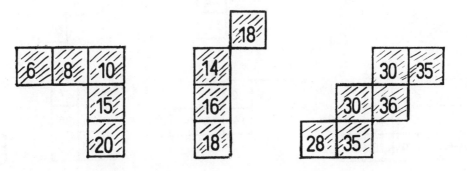

7. Find these same sections in a different position on the table.
   Shade the sections using a different shading.   Some shadings
   will overlap.

PSM 81

Multiplication Patterns

Mathematics teaching objective;

    . Practice multiplication skills.

Problem solving skills pupils <u>might</u> use:

    . Look for and use patterns.

    . Work backwards.

    . Search for or be aware of other solutions.

Materials needed:

    . None

Comments and suggestions:

. A similar activity could be developed using an addition table.

. The activity is a whole-group activity with the first page intended as a teacher-led discussion of patterns in the multiplication table. Pupils may need guidance on the second page. Occasionally some may need to refer to the completed table on the first page.

. Some pupils may count by multiples to fill in the blanks, e.g., 10, 15, 20,... . Others will work backwards to decide where the problem fits on the table. They can determine one factor by finding the difference between terms. For example, in | 4 | 8 | | the difference between terms is 4, meaning the factor at the far left of the table is 4 and the factors at the top are 1, 2 and 3. So the missing term is 12. (See the diagram at the right.)

Answers:

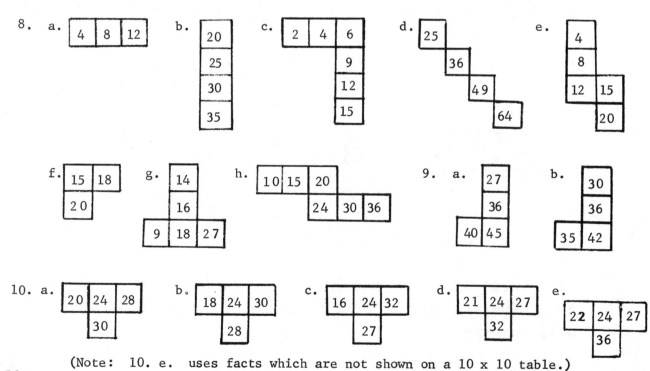

(Note: 10. e. uses facts which are not shown on a 10 x 10 table.)

Multiplication Patterns (cont.)

The following are sections from a multiplication table.

8. Fill in the missing numbers.   Try to do it without looking
at your completed table.

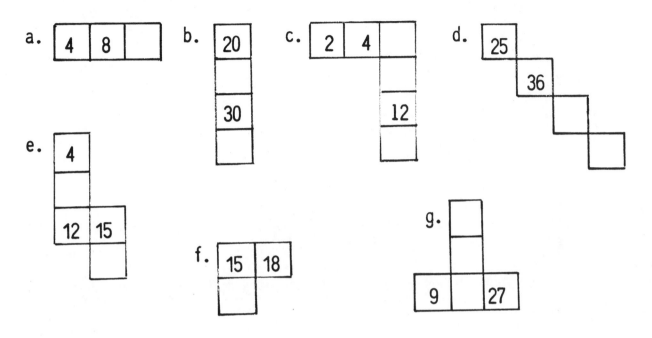

a. | 4 | 8 | |

b. | 20 |
   | |
   | 30 |
   | |

c. | 2 | 4 | |
   12

d. 25
   36

e. | 4 |
   | 12 | 15 |

f. | 15 | 18 |

g.
   | 9 | 27 |

h. | 10 | 15 | | |

9. Find two different
    ways to do this one.

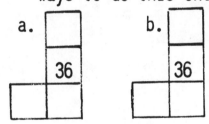

a.
   36

b.
   36

10. Find as many as you can for this one.
    Record your answers.

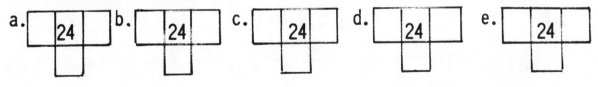

a. | 24 | b. | 24 | c. | 24 | d. | 24 | e. | 24 |

THE OUTERS AND THE INNERS

1. Carlos told his teacher about a discovery he had made. Ms. Harris thought others should try it. See if you can find the discovery.

   a. Use four consecutive numbers like 3, 4, 5, 6.

   b. Find the product of the inside numbers. 4 x 5 = 20

   c. Find the product of the outside numbers. 3 x 6 = 18

   d. Try other numbers.

Numbers	Inside Product	Outside Product
3, 4, 5, 6		
5, 6, 7, 8		
9, 10, 11, 12		
15, 16, 17, 18		
50, 51, 52, 53		

→ Make up your own numbers.

Discovery: _____

2. Debbie wondered about using four consecutive even numbers like 4, 6, 8, 10. Try five cases.

Discovery: _____

Numbers	Inside Product	Outside Product

3. Lonnie thought something different would happen if four consecutive odd numbers like 7, 9, 11, 13 were used. Try five cases.

Discovery: _____

Numbers	Inside Product	Outside Product

The Outers and the Inners

Mathematics teaching objectives:

. Develop mathematics vocabulary such as <u>product</u>, <u>even</u>, <u>odd</u>, <u>consecutive</u>.
. Practice multiplication skills.

Problem solving skills pupils <u>might</u> use:

. Make and use a table.
. Look for and use patterns.
. Create new problems by varying an old one.

Materials needed:

. Calculators (optional)

Comments and suggestions:

. The activity provides drill and practice in multiplication of whole numbers. A calculator could be used as motivation.
. Each pupil needs a copy of the activity. Introducing the activity by using a transparency may be helpful.

Answers:

1.

	Inside Product	Outside Product
3, 4, 5, 6	20	18
5, 6, 7, 8	42	40
9,10,11,12	110	108
15,16,17,18	272	270
50,51,52,53	2652	2650

Discovery: <u>The inside product is 2 greater than the outside product</u>.

2. For four even consecutive numbers, the inside product is 8 greater than the outside product.

3. For four odd consecutive numbers, the inside product is 8 greater than the outside product.

Extension:

Pupils could investigate other patterns such as the inside and outside products for four consecutive multiples of 3, 4, etc.

OPERATION, PLEASE

Study the examples in the first two columns. Write the operation in the blank. Complete the table. Create your own problems for the last two columns.

1.

Operation _____

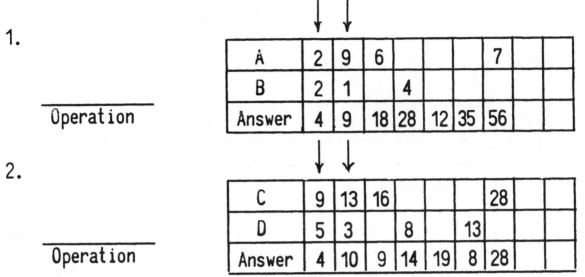

A	2	9	6				7		
B	2	1		4					
Answer	4	9	18	28	12	35	56		

2.

Operation _____

C	9	13	16				28		
D	5	3		8		13			
Answer	4	10	9	14	19	8	28		

3. Create a one-operation table of your own.

Operation _____

E									
F									
Answer									

4. One operation was used to get answer 1. A different operation was used to get answer 2. Discover both operations and complete the table.

Operation 1 _____

Operation 2 _____

G	7	5	11			7		
H	4	8						
Answer 1	11	13	19	22	12		6	
Answer 2	28	40			35	42	5	

5. Create a two-operation table of your own.

-61-

Operation, Please

Mathematics teaching objective:

. Practice computation skills.

Problem solving skills pupils <u>might</u> use:

. Identify patterns suggested by data in tables.

. Create new problems by varying an old one.

Materials needed:

. None

Comments and suggestions:

. Be sure pupils understand that each column represents a problem within a table. The last two columns are blank for pupils to create their own problems.

. If the operations of subtraction and division are chosen for the last problem, care is necessary to avoid negative and/or fractional answers.

Answers:

1.

Operation - <u>multiplication</u>

Answers may vary.

A	2	9	6	7			7		
B	2	1	3	4			8		
Answer	4	9	18	28	12	35	56		

2.

Operation - <u>subtraction</u>

Answers may vary.

C	9	13	16	22		21	28		
D	5	3	7	8		13	0		
Answer	4	10	9	14	19	8	28		

4.

Operation 1 - <u>addition</u>

Operation 2 - <u>multiplication</u>

Answers may vary.

G	7	5	11		5	7	1		
H	4	8	8		7	6	5		
Answer 1	11	13	19	22	12	13	6		
Answer 2	28	40	88		35	42	5		

SCORE WITH FOUR

Get 3 regular dice.
Get a playing board.
Get a partner.

Directions:

1. Decide who goes first.

2. First player

   . Roll the dice.

   . Use the numbers and 1 or 2 operations to make a number statement. (An example for 2, 4, and 5 is 2 x 5 + 4 = 14.)

   . Mark an "X" on the answer on the playing board.

3. Second player

   . Roll the dice.

   . Make a number statement.

   . Mark your answer with an "O".

4. Continue to take turns.

5. The first player to get 4 marks in a row wins.

Ⓒ PSM 81

Score With Four

Mathematics teaching objectives:

. Practice basic computational skills.

. Write or say number statements.

Problem-solving skills pupils might use:

. Work backwards.

. Guess and check.

. Record solution attempts.

Materials needed:

. 3 regular dice for every 2 pupils

Comments and suggestions:

. Introduce the game to the total class (a transparency is useful).

. Be certain all pupils understand the directions. Then have partners play the game.

. Since this game helps pupils develop their number sense, include it frequently when the goal of instruction is drill and practice through games.

. A large laminated playing board and differently colored markers eliminates the need to continually reproduce this sheet.

Answers:

Pupils should check the number statements as each partner has a turn. Some may write statements such as $3 + 4 \times 2 = 14$. Encourage the use of parentheses to show which operation is done first, $(3 + 4) \times 2 = 14$ or $3 + (4 \times 2) = 11$.

Score With Four (cont.)

PLAYING BOARD

1	2	3	4	5	6	7	8
9	10	11	12	13	14	15	16
17	18	19	20	21	22	23	24
25	26	27	28	29	30	31	32
33	34	35	36	37	38	39	40
41	42	44	45	48	50	54	55
60	64	66	72	75	80	90	96
100	108	120	125	144	150	180	216

PSM 81

CREATE A PROBLEM

Write a single digit in each square to create a correct problem.
Digits may be repeated in a problem.

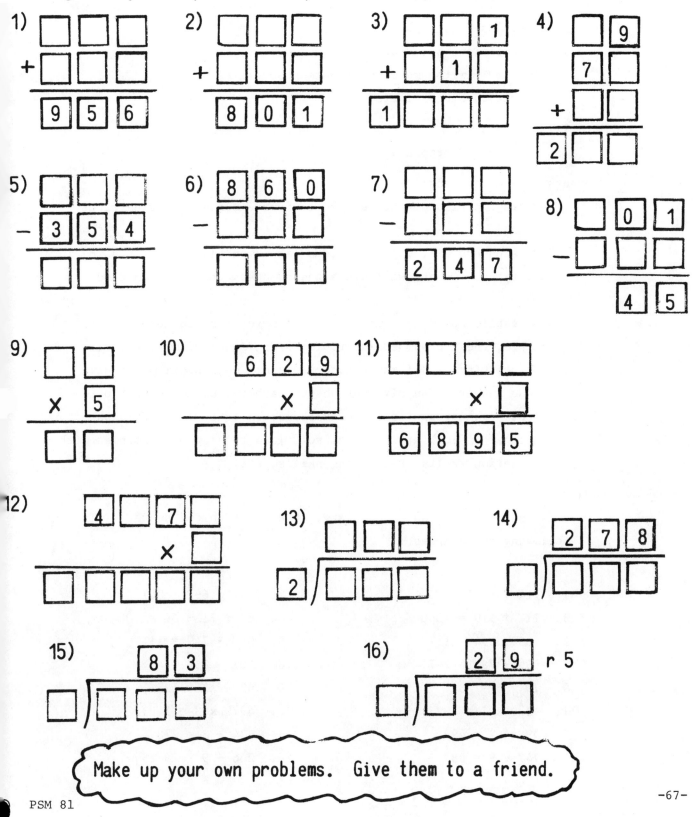

1)
$$\begin{array}{c}\square\ \square\ \square \\ +\ \square\ \square\ \square \\ \hline 9\ 5\ 6\end{array}$$

2)
$$\begin{array}{c}\square\ \square\ \square \\ +\ \square\ \square\ \square \\ \hline 8\ 0\ 1\end{array}$$

3)
$$\begin{array}{c}\square\ \square\ 1 \\ +\ \square\ 1\ \square \\ \hline 1\ \square\ \square\ \square\end{array}$$

4)
$$\begin{array}{c}\square\ 9 \\ 7\ \square \\ +\ \square\ \square \\ \hline 2\ \square\ \square\end{array}$$

5)
$$\begin{array}{c}\square\ \square\ \square \\ -\ 3\ 5\ 4 \\ \hline \square\ \square\ \square\end{array}$$

6)
$$\begin{array}{c}8\ 6\ 0 \\ -\ \square\ \square\ \square \\ \hline \square\ \square\ \square\end{array}$$

7)
$$\begin{array}{c}\square\ \square\ \square \\ -\ \square\ \square\ \square \\ \hline 2\ 4\ 7\end{array}$$

8)
$$\begin{array}{c}\square\ 0\ 1 \\ -\ \square\ \square\ \square \\ \hline 4\ 5\end{array}$$

9)
$$\begin{array}{c}\square\ \square \\ \times\ 5 \\ \hline \square\ \square\end{array}$$

10)
$$\begin{array}{c}6\ 2\ 9 \\ \times\ \square \\ \hline \square\ \square\ \square\ \square\end{array}$$

11)
$$\begin{array}{c}\square\ \square\ \square\ \square \\ \times\ \square \\ \hline 6\ 8\ 9\ 5\end{array}$$

12)
$$\begin{array}{c}4\ \square\ 7\ \square \\ \times\ \square \\ \hline \square\ \square\ \square\ \square\ \square\end{array}$$

13)
$$2\ \overline{)\ \square\ \square\ \square}\quad\text{with quotient }\square\ \square\ \square$$

14)
$$\square\ \overline{)\ 2\ 7\ 8}$$

15)
$$\square\ \overline{)\ \square\ \square\ \square}\quad\text{with quotient } 8\ 3$$

16)
$$\square\ \overline{)\ \square\ \square\ \square}\quad\text{with quotient } 2\ 9\ \text{r } 5$$

Make up your own problems. Give them to a friend.

PSM 81

Create a Problem

Mathematics teaching objectives:

  . Practice computation skills.

  . Use inverse operations.

Problem solving skills pupils might use:

  . Make decisions based upon data.

  . Recognize limits and/or eliminate possibilities.

  . Make reasonable estimates.

Materials needed:

  . Calculators (optional)

Comments and suggestions:

  . Encourage pupils not to use the obvious answers involving zeros
    and ones.

  . Some students will solve problems like 7 and 15 easily by using inverse
    operations. Others will use a guess, check and refine method.

  . Some pupils could determine the limits for certain numbers. For
    example, the top number in Problem 9 can range from 10 to 19.

  . This activity is difficult to score since each pupil might have
    unique answers. An alternative to teacher scoring is to provide
    calculators for pupils to check their own or a classmate's paper.
    You can then look at incorrect solutions to diagnose troubles with
    computation skills, regrouping skills, etc.

Answers:

  Answers for each problem can vary. None have unique answers. The
  following are samples.

  1. $111 + 845 = 956$ 2. $111 + 690 = 801$
  3. $111 + 919 = 1030$ 4. $69 + 71 + 60 = 200$
  5. $465 - 354 = 111$ 6. $860 - 749 = 111$
  7. $999 - 752 = 247$ 8. $901 - 856 = 45$
  9. $19 \times 5 = 95$ 10. $629 \times 2 = 1256$
  11. $1379 \times 5 = 6895$ 12. $4979 \times 3 = 14937$
  13. $998 \div 2 = 499$ 14. $834 \div 3 = 278$
  15. $747 \div 9 = 83$ 16. $266 \div 9 = 29 \text{ r}5$

THE LARGEST NUMBER GAME

1. Any number of pupils can play.

2. Get the digit cards (0 – 9).

3. Shuffle the cards. Draw 4 cards.

4. Fill the numbers in the blanks.

    ___ ___ ___ ___

5. Arrange the numbers in any order. Use +, –, x, ÷
    to make the largest <u>one-digit</u> answer.

EQUATION:

> Example: 3, 8, 1, 2
> $\underline{8} + \underline{3} - \underline{2} \div \underline{1} = \underline{9}$

   a. ___ ___ ___ ___ = ___

6. Repeat this four more times. Reshuffle the cards
    each time.

   b. ___ ___ ___ ___ = ___

   c. ___ ___ ___ ___ = ___

   d. ___ ___ ___ ___ = ___

   e. ___ ___ ___ ___ = ___

7. Use all five of your answers (a, b, c, d, and e). The
    pupil who arranges them to make the largest five-digit
    number wins the round.

    Largest number = ___ ___ , ___ ___ ___

The Largest Number Game

Mathematics teaching objectives:
- . Practice computation skills.
- . Develop understanding of identity elements, 0 and 1.

Problem-solving skills pupils *might* use:
- . Translate a problem into mathematical notation.
- . Make decisions based upon data.
- . Guess and check.

Materials needed:
- . Digit cards 0-9

Comments and suggestions:
- . This activity is best used with small groups but can be done with the whole class with the teacher drawing the four digit cards. For a particular round, all pupils use the same four digits.
- . Pupils will try to solve the problem using as few operations as possible. Addition and subtraction will be used most--division the least except for division by 1.
- . A reasonable time limit per round may be needed. Point out that the 5 answers may be arranged in any order for the final answer.
- . Most pupils will write and perform the operations from left to right. If an explanation yields the appropriate answer, accept it. This activity may provide motivation for later study of the rules for order of operations and parentheses.
- . To avoid excess duplication of this page, have pupils draw their own answer blanks or laminate several copies which can be wiped off after use.

Answers:

Answers will vary for each problem according to the digits drawn.

Grade 5

III.   STORY PROBLEMS

III. STORY PROBLEMS

Story problems are an important part of problem solving in mathematics. They are used throughout mathematics and science courses as applications for concepts and skills. Their purpose is to help students transfer what they are learning in the classroom to everyday life and later to careers.

The teaching of story problems has been criticized in several ways. One complaint is that teaching story problems often is the <u>only</u> attempt to teach problem solving. As this book shows, problem solving is much more than just story problems. Another criticism is that a page of story problems may use only one skill, such as subtraction. In this case, students can decide what to do to solve problem 1 and avoid reading or understanding the rest. A third criticism is that students are not helped to <u>understand</u> the story problems but are unintentionally encouraged to race to the answers. These things can lead students to develop their own <u>incorrect rules</u> as Joe Dodson, Mathematics Supervisor for the Winston-Salem/Forsyth County Schools, illustrates in the <u>North Carolina State Math Newsletter</u>.

"A STUDENT'S GUIDE TO PROBLEM SOLVING (A Spoof)

"Rule 1: If at all possible, avoid reading the problem. Reading the problem only consumes time and causes confusion.

"Rule 2: Extract the numbers from the problem in the order in which they appear. Be on the watch for numbers written in words.

"Rule 3: If rule 2 yields three or more numbers, the best bet for getting the answer is adding them together.

"Rule 4: If there are only two numbers which are approximately the same size, then subtraction should give the best results.

"Rule 5: If there are only two numbers in the problem and one is much smaller than the other, then divide if it goes evenly-- otherwise, multiply.

"Rule 6: If the problem seems like it calls for a formula, pick a formula that has enough letters to use all the numbers given in the problem.

"Rule 7: If the rules 1-6 don't seem to work, make one last desperate attempt. Take the set of numbers found by rule 2 and perform about two pages of random operations using these numbers. You should circle about five or six answers on each page just in case one of them happens to be the answer. You might get some partial credit for trying hard.

"Rule 8: Never, never spend too much time solving problems. This set of rules will get you through even the longest assignments in no more than ten minutes with very little thinking."

The criticism listed previously can be avoided by teaching problem solving in many ways by providing sets of problems requiring several different skills and by giving activities where the objective is to understand the story problems, not "race to the answer." These ideas are incorporated in this section.

The six activities in this section use whole numbers and money. All four operations are involved. Although estimation is a main point of two activities, pupils should acquire the habit of checking the reasonableness of each answer. Three activities include ideas not often found in texts:

. pupils supply missing information needed to solve a problem.

. pupils identify extra information in a problem.

. pupils create problems to match given solutions.

VERY SHORT STORIES

Solve these problems, if possible.

> Marcia bought 7 records. How much did they cost her?

> Jose bought a pair of shoes for $14.95. How much change did he get back?

> Julie spent $1.50 for pencils. How many did she buy?

1. Read the above problems.

   a. What information is needed to solve each of them?

   b. Make up information. Then solve each problem.

2. Make up information for each problem. Then solve the problem.

   a. Sam has a part-time job. He earns $3.50 per hour. How much did he make last week?

   b. Cindy caught a 30 cm trout. How much longer was the trout that Ellen caught?

   c. Bananas are 35¢ per pound; apples are 43¢ per pound. How much did Mr. Jones spend?

   d. The odometer now reads 10,349 miles. How many miles did the Greens travel on their trip?

   e. Susan and Sarah averaged 20 kilometres per hour on their bicycle trip. How far did they go?

   f. Ross shared his cookies equally with 3 others. How many cookies did each of them get?

   g. Sharon bought a new tennis racket for $9.50. Later she sold it at a garage sale. How much did she lose?

   h. Big sale on fruit! Apples are 39¢ per pound; oranges are 10¢ each. How much for 8 pounds of apples and 5 pounds of oranges?

   i. Miss Hill spent $6.90 for some hamburgers and colas. If she bought 8 colas, how many hamburgers did she buy?

-75-

Very <u>Short Stories</u>

Mathematics teaching objectives:

- Solve single step word problems.
- Compute with money amounts.
- Practice basic fact operations.

Problem solving skills pupils <u>might</u> use:

- Create data needed to solve a problem.
- Use math symbols to translate and solve a problem.

Materials needed:

- None

Comments and suggestions:

- This page prevents students from racing to an answer. They must first decide what is missing. Here is a chance to emphasize how important it is to understand the problem. Students can be given credit for identifying what's missing <u>before</u> they are asked to find the answers. <u>In fact, you might not have them find the answers at all for this activity</u>.
- Using the overhead projector or a sheet for each pupil, allow them to ponder and struggle with the first examples. Discuss the needed information and then solve the problems in the bubbles.
- Encourage pupils to use reasonable information in completing the activity. You may want pupils to write a sentence describing the needed information.
- Many pupils will think that problem <u>h</u> can be solved as is!
- Students could switch papers to check computation by hand or calculator.

Answers:

1. a. What is the cost of each record?
      How much did Jose pay the clerk?
      What is the cost of a pencil?

2. a. How many hours did Sam work?
   b. What is the length of Ellen's trout?
   c. What amounts of bananas and apples did Mr. Jones buy?
   d. What was the odometer reading at the end of the trip?
   e. How many hours did Susan and Sarah ride?
   f. How many cookies does Ross have?
   g. For what amount did Sharon sell her racket?
   h. How many oranges were bought? or   What is the cost of a pound of oranges?
   i. What is the cost of a cola and a hamburger?

NUMBERS PLEASE

Amy counted ____ students in her class. Each of them brought in ____ cupcakes for a sale. How many cupcakes did they have to sell?

Numbers are needed to solve this problem. Decide what numbers will be <u>reasonable</u>. Then solve the problem above and the ones below.

1. Slim weighs ____ pounds. His father weighs ____ pounds. How much heavier is his father?

2. Kim bought ____ erasers for ____ cents each. How much did she spend?

3. James bought a pair of shoes for ____ and a pair of socks for ____ . How much did he spend?

4. Melody bought a box containing ____ pencils for ____ . How much did each pencil cost her?

5. Sid sold ____ magazines. Sandy sold ____ magazines. How many more did Sandy sell?

6. Tammy's father pays her to run errands. On Monday she made ____ and on Wednesday she made ____ . How much did she make?

7. Kenny had ____ gumdrops. He shared them equally with ____ friends. How many did each get?

8. Ginny counted ____ words on the first line of a page in her book. The page has ____ lines. About how many words are on that page?

9. Ross gets an allowance of ____ per week. Last year it was ____ per week. How much more per week is he getting this year?

10. Sheila is ____ cm tall. Jill is ____ cm tall. Gary is ____ tall. What is their average height?

11. During the summer Sammy sells lemonade for ____ cents a glass. One day he sold ____ glasses. How much did he make?

12. Jeff was given ____ tiles. He was asked to arrange them so they would form a square. How many does he put on each side?

PSM 81

<u>Numbers</u> <u>Please</u>

Mathematics teaching objectives:

    . Solve single step word problems.

    . Compute with money amounts.

    . Practice basic fact operations.

Problem solving skills pupils <u>might</u> use:

    . Simplify the problem.

    . Make reasonable estimates.

    . Use math symbols to translate and solve a problem.

Materials needed:

    . None

Comments and suggestions:

    . Pupils can use very small numbers to determine the necessary operation. Then, reasonable numbers for each item can be determined and the problem can be solved. This is a very useful problem solving strategy that can be pointed out: Simplify the problem.

    . The activity provides an opportunity to diagnose pupil recognition of money values and sizes of some "real life" items. Pupils may lack a feeling for common everyday costs.

    . A discussion of the following items may be needed:

        . Problem 8 calls for an approximate answer.

        . Problem 10 involves finding an average.

        . Problem 12 requires pupils to know the number of tiles needed to make a square such as 9, 16, 25, etc. Some pupils may create squares using the tiles just around the outside. Pupils should have the idea that looking at a problem in different ways is o.k.   12 tiles

    . Students could switch papers to check computation by hand hand or calculator.

Answers:

    Answers for this activity will vary according to the numbers chosen by each pupil.

WRITING SHORT STORIES

Calvin got 100% on his story problem lesson.
His work is shown below.
Can you think of story problems that match his work?
Write your problems in the space provided.

CALVIN'S WORK	YOUR STORY PROBLEMS
① 73 +29 My answer is 102 newspapers.	
② 9 × $1.98 My answer is $17.82.	
③ $100 − $73.95 My answer is $26.05.	
④ 253 −194 My answer is 59 miles.	
⑤ 7⟌$8.54 My answer is $1.22.	
⑥ $.29 .49 + .39 My answer is $1.17.	
⑦ 5⟌865 The average weight is 173 pounds.	
⑧ 25 ×12 My answer is 300 eggs.	
⑨ 3 × $.59 + 5 × $.75 My answer is $5.52.	
⑩ 8⟌40 My answer is 5 cookies.	

Writing Short Stories

Mathematics teaching objective:

. Create single and multiple step word problems.

Problem solving skills pupils <u>might</u> use:

. Search printed material for needed information.

. Create a problem which has been solved by a set of calculations.

Materials needed:

. None

Comments and suggestions:

. Students are often asked to translate "real life" situations into math symbols. Here is an opportunity to do the reverse. Again, the focus is on understanding the problem.

. The activity provides each pupil a chance to express his/her creativity. Some will create the minimum amount while others will produce an intricate problem often involving unnecessary information.

. Problem 7 involves an average weight which may need to be explained. For those pupils not understanding the concept of average, you could use this problem as an example.

> Five large bags of potatoes hold 160, 185, 182, 178 and 170 pounds of potatoes. Sally has the job of moving potatoes from bag to bag until each bag has the same number of pounds of potatoes. How many pounds will each bag have?

Answers:

Answers will vary.

ESTIMATING

Gretchen is helping her father do the shopping. They already have five items in their cart.

Eggs	$ .89	Laundry soap	$1.99
Cookies	1.09	Milk	1.59
Roast	4.05		

Do you think that $10 will be enough? Estimate to see.

1. Four grocery items are $3.09, $1.89, $.49, and $2.61. About how much is the total?

    a. $5.00   b. $6.00   c. $7.00   d. $8.00

2. Grace bought 16 pens at 39¢ each. About how much did she spend?

    a. Less than $3.00   b. Between $3.00 and $4.00

                  c. More than $4.00

3. During 1976 the O'Briens spent $628 on gas. Then, they bought a smaller car. Their gas bill for 1977 dropped to $379. About how much did they save?

    a. $1,000   b. $350   c. $250   d. $300

4. Big Spender bought 4 TVs at $498.95 each and 3 stereos at $189.50 each. About how much did he spend?

    a. $2,500   b. $3,000   c. $500   d. $700

5. Julie's Girl Scout troop returned 18 cartons of empty bottles to the store. Each carton contained 24 bottles. About how many bottles did they return?

    a. 5,000   b. 500   c. 50   d. 1,000

6. In five days, Alice rode 79 km, 86 km, 49 km, 114 km, and 79 km. About how far did she ride altogether?

    a. 500 km   b. 450 km   c. 400 km   d. 300 km

Estimating

Mathematics teaching objectives:

. Solve single step word problems by estimating.

. Compute with money amounts.

. Use mental calculations.

Problem solving skills pupils <u>might</u> use:

. Make reasonable estimates as answers.

. Use math symbols to translate and solve a problem.

Materials needed:

. None

Comments and suggestions:

. Often students are more comfortable doing exact computations than they are doing estimating. Practice in estimation can help overcome this resistance.

. After all estimates have been made, pupils may want to find exact answers to see how close they were to the estimates. Calculators could be used. If students are <u>required</u> to compute the exact answers, they may be less motivated to estimate in the first place.

. Emphasize that an estimate is seldom wrong--some estimates are just closer than others.

Answers:

1. $8.00
2. More than $4.00
3. $250
4. $2,500
5. 500
6. 400 km
7. 700
8. $700
9. 30
10. A little less than 5 hours
11. 20 months
12. $6.00 or $6.60
13. 1800 words
14. 1800 miles

Estimating (cont.)

7. About how large is the total enrollment at Westside School?

1st grade	123	4th grade	131
2nd grade	81	5th grade	147
3rd grade	109	6th grade	141

    a. 400   b. 500   c. 600   d. 700

8. Big Spender's bill is $289.95. About how much change should he get back from a $1,000 bill?

    a. $700   b. $800

    c. $810   d. $1,300

9. Stamps cost 15¢ each. About how many can Joe buy for $5.00?

    a. 10   b. 20   c. 30   d. 40

10. On the average, Mr. Keene can read a little over one page per minute. About how long will it take him to read 300 pages?

    a. A little less than 5 hours   c. 3 hours
    b. A little more than 5 hours   d. 10 hours

Make reasonable estimates for each of the following.

11. A TV costs $398.95. Payments are $19.95 per month. About how many months will it take to pay for the TV?

12. Miss Thrifts bought 11 gallons of Super-Gas at 89.9¢ per gallon. About how much did she spend?

13. Evan can type 29 words in a minute. About how many words is this per hour?

14. It is 2,818 miles from Portland to Washington, D.C. It is 980 miles from Portland to Los Angeles. From Portland, about how much further is it to Washington, D.C. than to Los Angeles?

WHAT'S NOT NEEDED?

Read the following problems carefully. Decide what information is not necessary and cross it out. Solve the problem.

1. Rose weighs 85 lbs. She weighed 15 lbs. less last year. Her father weighs 176 lbs. How much more does he weigh? _____

2. John ran a quarter-mile in 81 seconds. His sister ran it in 75 seconds. John ran the quarter-mile last year 18 seconds slower. What was his time last year?

3. Susan left on a trip from New York to San Francisco. The odometer was 88,761 when she started. The odometer was 90,172 in Phoenix, and 90,935 in San Francisco. How far was her journey?

   8 8 7 6 1 0

   9 0 1 7 2 8

   9 0 9 3 5 7

4. Joe's Donut Shop sells butterhorns for 15¢ each and maple bars for 20¢. How much would one dozen maple bars cost?

5. Jill's parents bought a house in 1974 for $24,500. They made monthly payments of $185. They sold the house for $31,750 in 1977. How much did they pay on the house in one year?

6. Mike went shopping and bought the following items: a baseball glove for $12.95, a shirt for $8.98, a pair of blue jeans for $12.50, and a yo-yo for $1.95. How much did he spend on clothes?

7. A Boeing 707 cruises at 450 miles per hour but is capable of speeds up to 600 miles per hour. A 747 cruises at 600 miles per hour but may do 725 miles per hour for short periods. How much faster is the 747's cruising speed?

8. In Portland, 895 people ran the Marathon--120 females and 775 males. 729 people finished. How many did not finish?

9. Make up a problem of your own that has too much information. Give it to a classmate to solve.

What's Not Needed?

Mathematics teaching objectives:

- Solve single step word problems.
- Compute with money amounts.
- Practice basic fact operations.

Problem solving skills pupils _might_ use:

- Eliminate (or ignore) data not needed.
- Use math symbols to translate and solve a problem.

Materials needed:

- None

Comments and suggestions:

- Most problems outside the mathematics text contain more information than is needed. This activity focuses on this attribute of problems. Pupils will need to read each problem carefully, locate the unnecessary information and cross it out. To focus students' attention on the importance of understanding the problem, credit can be given for crossing out the extra information before students finish solving the problem.
- A collection of pupil created problems could be dittoed and used as a follow-up to this activity.

Answers:

	Unneeded Information	Answer
1.	She weighed 15 pounds less last year.	91 pounds
2.	His sister ran it in 75 seconds	99 seconds
3.	The odometer was 90,172 when she was in Phoenix	2174 miles
4.	butterhorns for 15¢ each	$2.40
5.	They sold the house for $31,750 in 1977.	$2220
6.	a baseball glove for $12.95, a yo-yo for $1.95	$21.48
7.	speeds up to 600 miles per hour.... may do 725 miles per hour	150 miles per hr
8.	120 females and 775 males.	166 people

THE HIDDEN QUESTION

Find the answer to these problems.

James bought a hamburger for 89¢ and a cola for 39¢. He gave the clerk a $5.00 bill. How much change did he get?

How did you go about solving this problem?

What "hidden question" did you have to answer before you could determine the amount of change?

Each problem below has a hidden question. First answer the hidden question. Then find the answer to the problem.

1. Martha bought a used record for $1.89 and another for $2.98. She gave the clerk a $10.00 bill. How much change did she get?

   Answer to hidden question: _____

   Final answer: _____

2. Jim is a busboy. He makes $15.00 per day plus tips. Last week he made $27.50 in tips. What were his total earnings during those five days?

   Answer to hidden question: _____

   Final answer: _____

3. At the Tasty Donut Shop, donuts sell for $2.40 per dozen. At this rate, how much will 14 donuts cost?

   Answer to hidden question: _____

   Final answer: _____

4. Sue, Sandy, and Sarah bought presents costing 98¢, 99¢, $1.19, and $1.49. They want to share the cost. How much should each of them pay?

   Answer to hidden question: _____

   Final answer: _____

The Hidden Question

Mathematics teaching objectives:

. Solve multiple step word problems

. Compute with money amounts.

Problem solving skills pupils might use:

. Break a problem into manageable parts.

. Use math symbols to translate and solve a problem.

Materials needed:

. None

Comments and suggestions:

. Show pupils an example of this type of problem. They probably will
be able to find the answer to the hidden question but may not be
able to verbalize it. Encourage them to write down the hidden question.

Answers:

1.	How much for both records?	$ 4.87	$ 5.13
2.	What were his wages for 5 days?	$75.00	$102.50
3.	How much for one donut?	$ .20	$ 2.80
4. .	How much for the presents?	$ 4.65	$ 1.55
5.	How many cookies in all?	96	8 dozen
6.	How many feet in two pieces?	7 feet	4 feet
7.	How much for 5 hamburgers?	$ 4.00	4 cokes
8.	What is the monthly cost for a year?	$ 9.00	$ 3.00

The Hidden Question (cont.)

5. Mrs. Green is buying cookies for the class party. She expects that each pupil will eat 4 cookies. How many dozen will she need if there are 24 pupils in the class?

Answer to hidden question: _____

Final answer: _____

6. Jean needs to cut a board into two $3\frac{1}{2}$-foot pieces. She starts with an 11-foot board. How much of the board will she have left after she makes the cuts?

Answer to hidden question: _____

Final answer: _____

7. Slim has $6.00 to spend on hamburgers and colas. Hamburgers cost 80¢ each and colas cost 50¢ each. If he buys 5 hamburgers, how many colas can he buy?

Answer to hidden question: _____

Final answer: _____

8. A monthly magazine sells for 75¢ per month at the store. If you buy a yearly subscription it costs $6.00. How much do you save by buying the yearly subscription?

Answer to hidden question: _____

Final answer: _____

Grade 5

IV. FRACTIONS

IV. FRACTIONS

It seems perfectly sensible to many
pupils to add fractions straight across.
They have forgotten, or did not learn,
the meaning for fraction addition. When
asked the answer for 2 apples plus 3
apples, most will answer 5 apples. When
asked the answer for 2 tenths plus
3 tenths (in words), pupils usually
answer 5 tenths. But when asked to do
$\frac{2}{10} + \frac{3}{10}$, many will answer $\frac{5}{20}$. In
words they do fine but in symbols
they make up rules.

Pupils need much work with meanings of symbols and operations before
they can be expected to use the symbols and operations with understanding.
They need to be experienced in working with a model so they can return
again and again to the model when they forget what to do. Research* indicates
that the region model, using a circle or rectangle as a whole, should be
mastered before a set model or number line model for fractions is used. The
fifth-grade seems an appropriate place for extensive work with the region
model.

The activities in this section use circles and pieces of circles to
build background for fraction concepts and operations. Word names like
3 tenths are emphasized along with symbolic names like $\frac{3}{10}$. No work is
done with formal algorithms although the activities could be used to develop
algorithms for finding equivalent fractions, sums, differences or products.
This section is very developmental. As the concepts are learned the stage
is set for more problem solving with fractions in later grades.

Using the Activities

The amount of this unit you do will depend upon your course of study.
The activities are sequenced from concepts of naming fractions through the

*As discussed in DEVELOPING COMPUTATIONAL SKILLS, NCTM 1978 Yearbook.

concept of division. Each lesson is designed to make full use of the fraction pieces as a manipulative to better understand the concepts.

A time is eventually reached when pupils will not feel the need to use the fraction pieces. As one fifth-grade girl said, "I'll use them when I need them." You as a teacher should encourage use of the pieces. At any time during an activity, pupils should be able to demonstrate the solution to a problem.

Special note: The emphasis in this section is on understanding the operation. No work with formal algorithms is done. You will need to go to the textbook for the algorithms. You will find, though, that many activities can be used to develop the algorithm. For example, $\frac{1}{2} + \frac{1}{3}$ is done by covering the $\frac{1}{2}$ piece with three $\frac{1}{6}$ pieces and by covering the $\frac{1}{3}$ piece with two $\frac{1}{6}$ pieces to show $\frac{1}{2} + \frac{1}{3} = \frac{3}{6} + \frac{2}{6} = \frac{5}{6}$.

Because of slight inaccuracies in cutting and because of the smallness of several pieces like the tenths and twelfths, pupils will sometimes get false results. What is called an "exact covering" may be slightly off. Having pupils work together when answers differ is a partial solution to this problem.

Black masters of the circle pieces are provided. A thermofax master can be used to ditto the circles onto construction paper. Four circles can be placed on each sheet so eight sheets, per color, are needed for a class of 32 pupils. The circles can be cut out by an aide or by the pupils themselves. If class time is used, about 30 minutes will be needed. An envelope is needed for each pupil to store the pieces. It's helpful to have several spare pieces to replace those that get lost. The pieces need to be made only once and then can be saved for year-to-year use.

A set of transparent circle fraction pieces is very helpful for teacher demonstrations on the overhead projector. Colored pieces can be produced using the Diazo process if such a machine is available. Or pieces can be colored with felt pens.

WHAT'S MY NAME?

Needed: Circle fraction pieces

1.   How many different colors are there?   _____

2.   What is the color of the smallest piece?   _____

                              largest piece?   _____

3.   Arrange one piece of each color in order from <u>smallest</u> to <u>largest</u>.
     Record the order here. (Use color names.)

     _____ _____ _____ _____ _____ _____ _____ _____ _____

4.   Which color shows <u>1 whole</u> circle?   _____

5.   Show how all the green pieces exactly cover the <u>1 whole</u> piece.

6.   Write the number of green pieces needed to exactly cover 1 whole.____

7.   Do 5 and 6 with each of the colors.

     ____, ____, ____, ____, ____, ____, ____

8.   The white piece is <u>1 whole</u>. Which color shows:

     a. 1 half?   _____   c. 1 twelfth?   _____

     b. 1 sixth?   _____   d. 1 fifth?   _____

9.   Write the word name, like <u>1 third</u>, for 1 piece of these colors.

     a. 1 green   _____   c. 1 light blue   _____

     b. 1 red   _____   d. 1 dark blue   _____

10. Write the fraction name for these.

     a. 5 orange   _____   c. 2 purple   _____

     b. 3 red   _____   d. 3 yellow   _____

<u>What's My Name</u>?

Mathematics teaching objectives:
- . Name parts of a circle as fractions.
- . Relate word names to fraction notation.

Problem-solving skills pupils <u>might</u> use:
- . Use a model.
- . Make and use a table.

Materials needed:
- . Circle fraction pieces.

Comments and suggestions:
- . This activity familiarizes pupils with colors, sizes and names of the circle fraction pieces. It is important that pupils cover the white piece to show that all the pieces of a color <u>do</u> make one whole circle.
- . A list of color names and fraction words (half, third, fourth, ...) will be useful as reference for spelling.
- . After the tables are completed, pair pupils up. Have one pupil close his/her eyes while the other pupil hands him/her a fraction piece. By touch, the first pupil tries to decide what the piece is. It's best to use only $\frac{1}{5}$, $\frac{1}{4}$, $\frac{1}{3}$, $\frac{1}{2}$ or the whole. Pupils then reverse roles.
- . <u>Pupils</u> <u>may</u> <u>need</u> <u>the</u> <u>completed</u> <u>table</u> <u>in</u> <u>later</u> <u>activities</u>.
- . This activity takes about 25 minutes to complete.

Answers:

1. 9   2. orange, white   3. orange, red, dark blue, pink, purple, light blue, green, yellow, white.

4. white   8. a. yellow   b. pink   c. orange   d. purple

9. a. one-third   b. one-tenth   c. one-fourth   d. one-eighth

10. a. $\frac{5}{12}$ b. $\frac{3}{10}$ c. $\frac{2}{5}$ d. $\frac{3}{2}$

11. a. four-fourths   b. two-thirds   c. five-sixths   d. eleven-eighths

What's My Name (cont.)

11. Write the word name for these.

a. 4 light blue _____   c. 5 pink _____

b. 2 green _____   d. 11 dark blue _____

12. Jane developed a table to show all the fraction pieces.
Study her system. Complete the table.

Color	Number of Pieces To Make 1 Whole	Fraction Name For 1 Piece	Word Name For 1 Piece
White		////////	1 whole
Yellow			1 half
Green			
Light blue		$\frac{1}{4}$	
Purple			
Pink			
Dark blue			1 eighth
Red			
Orange		$\frac{1}{12}$	

-97-

What's My Name ?

Color	Number of Pieces To Make 1 Whole	Fraction Name For 1 Piece	Word Name For 1 Piece
White	1	/////////	1 whole
Yellow	2	$\frac{1}{2}$	1 half
Green	3	$\frac{1}{3}$	1 third
Light Blue	4	$\frac{1}{4}$	1 fourth
Purple	5	$\frac{1}{5}$	1 fifth
Pink	6	$\frac{1}{6}$	1 sixth
Dark Blue	8	$\frac{1}{8}$	1 eighth
Red	10	$\frac{1}{10}$	1 tenth
Orange	12	$\frac{1}{12}$	1 twelfth

WHICH IS LARGER?

Needed: Circle fraction pieces

1. Use one piece of each color. Lay one on top of the other to decide which is larger. Circle the <u>larger</u> one.

   a. 1 green or 1 yellow

   b. 1 purple or 1 pink

   c. 1 third or 1 fourth

   d. 1 twelfth or 1 tenth

   e. $\frac{1}{8}$ or $\frac{1}{6}$

   f. $\frac{1}{5}$ or $\frac{1}{10}$

2. Use the fraction pieces to decide. Circle the <u>larger</u> one.

   a. 2 thirds or 3 fourths

   b. 4 sixths or 4 fifths

   c. 1 whole or 2 thirds

   d. $\frac{3}{8}$ or $\frac{1}{4}$

   e. $\frac{2}{3}$ or $\frac{7}{8}$

   f. $\frac{4}{8}$ or $\frac{1}{2}$

3. Exactly cover each of these with twelfths pieces. Circle the <u>larger</u> one.

   a. $\frac{1}{6}$ or $\frac{1}{3}$

   b. $\frac{2}{3}$ or $\frac{2}{4}$

   c. $\frac{3}{6}$ or $\frac{1}{3}$

4. Suppose you had many other circle fraction pieces. Decide which is <u>larger</u> and circle it.

   a. 1 seventh or 1 eighth

   b. 3 ninths or 3 eighths

   c. $\frac{6}{7}$ or $\frac{6}{9}$

   d. $\frac{10}{11}$ or 1

5. Use the fraction pieces. Circle the larger one. Use the fraction pieces to decide how much larger it is.

   a. $\frac{1}{2}$ or $\frac{2}{3}$ _____

   b. $\frac{1}{2}$ or $\frac{2}{5}$ _____

   c. $\frac{2}{6}$ or $\frac{1}{4}$ _____

Which Is Larger?

Mathematics teaching objectives:

    . Compare fractions using parts of circles.

    . Compare unlike fractions by changing to like fractions.

Problem-solving skills pupils might use:

    . Use a model.

    . Recognize properties of an object.

Materials needed:

    . Circle fraction pieces

Comments and suggestions:

    . Have pupils work in pairs. One pupil closes his/her eyes while the
      other hands him/her two fraction pieces. By touch the first pupil
      decides which of the two pieces is larger (or perhaps the same). Pupils
      then reverse roles.

    . Emphasize actual covering of the pieces while pupils work the sheet.
      The covering dramatically demonstrates which is larger or smaller.

    . Problem 3 needs a teacher demonstration. It shows, informally, that
      two fractions can be compared by changing each to fractions with the
      same denominator.

    . Lining up one piece of each color and leaving blank spaces for missing
      fractions may help for problem 4.

    . This activity takes about 30 minutes to complete.

Answers:

    1. a. 1 yellow   b. 1 purple   c. 1 third   d. 1 tenth   e. $\frac{1}{6}$ f. $\frac{1}{5}$

    2. a. 3 fourths   b. 4 fifths   c. 1 whole   d. $\frac{3}{8}$ e. $\frac{7}{8}$ f. equal

    3. a. $\frac{1}{3}$ b. $\frac{2}{3}$ c. $\frac{3}{6}$

    4. a. $\frac{1}{7}$ b. $\frac{3}{8}$ c. $\frac{6}{7}$ d. 1

    5. a. $\frac{2}{3}$, $\frac{1}{6}$ larger   b. $\frac{1}{2}$, $\frac{1}{10}$ larger   c. $\frac{2}{6}$, $\frac{1}{12}$ larger

LARGER OR SMALLER THAN 1 ?

Needed: Two pupils
         Two sets of circle fraction pieces

1. Look at the fraction pieces. Predict whether each of the problems below will be (a) <u>more</u> than 1 whole, (b) <u>less</u> than 1 whole, or (c) <u>equal</u> to 1 whole.

    a. 3 yellow   _____   d. 12 red   _____

    b. 7 purple   _____   e. 4 light blue _____

    c. 6 dark blue   _____   f. 5 pink   _____

    Check your predictions.

2. Use the fraction pieces. How many

    a. $\frac{1}{5}$ pieces are needed to make 1 whole and 4 fifths? _____fifths

    b. $\frac{1}{8}$ pieces are needed to make 1 whole and 3 eighths? _____eighths

    c. $\frac{1}{10}$ pieces are needed to make $1\frac{2}{10}$? _____tenths

    d. $\frac{1}{4}$ pieces are needed to make $2\frac{1}{4}$? _____fourths

3. Use the fraction pieces. How much larger than 1 whole is

    a. 7 sixths? _____   c. $\frac{4}{3}$? _____

    b. 15 twelfths? _____   d. $\frac{9}{5}$? _____

Larger or Smaller Than 1?

Mathematics teaching objectives:

. Understand concept of converting from mixed to improper fractions.

. Understand concept of converting from improper to mixed fractions.

Problem-solving skills pupils <u>might</u> use:

. Use a model.

. Make reasonable estimates.

. Look for and use patterns.

Materials needed:

. Circle fraction pieces

Comments and suggestions:

. If you have transparent fraction pieces, begin this activity by
   showing a fraction piece and asking questions like:

    a. Will 3 (or whatever) of these pieces be more than, less
   than or equal to 1 whole circle?

    b. How many pieces are needed to make 1 (or 2) whole circles?

. Pupils need to work in pairs to show, for example, that 7 purple
   pieces are more than 1 whole circle. Again it's important that
   the pieces be used to demonstrate this concept.

. Pupils will need help interpreting $1\frac{2}{10}$ as 1 whole and 2 tenths.

. This activity takes about 20 minutes.

Answers:

1.   a. more   b. more   c. less   d. more   e. equal   f. less

2.   a. nine-fifths   b. eleven-eighths   c. twelve-tenths   d. nine-fourths

3.   a. 1 sixth   b. 3 twelfths   c. 1 third   d. 4 fifths

4.   a. 12   b. 13   c. 30

5.   a. $\frac{4}{13}$ b. $\frac{3}{11}$ c. $\frac{5}{14}$

Larger or Smaller Than 1 ? (cont.)

Suppose you had other circle fraction pieces.

4. a. How many $\frac{1}{7}$ pieces are needed to make $1\frac{5}{7}$? _____

    b. How many $\frac{1}{9}$ pieces are needed to make $1\frac{4}{9}$? _____

    c. How many $\frac{1}{15}$ pieces are needed to make 2 wholes? _____

5. How much larger than 1 whole is

    a. $\frac{17}{13}$? _____   b. $\frac{14}{11}$? _____   c. $\frac{19}{14}$? _____

ANOTHER NAME

Needed: Circle fraction pieces

1. Use a 1 half piece. Cover it <u>exactly</u> with pieces of another color.

    Example: 6 twelfths covers 1 half exactly.

    Find four more examples.

    _____   _____   _____   _____

2. Which fraction pieces cannot be used to exactly cover the 1 half piece? _____ _____

3. The exact coverings in (1) can be recorded like this:

    $\frac{1}{2} = \frac{6}{12}$ Record the other coverings you did.

    $\frac{1}{2} =$ _____ $\frac{1}{2} =$ _____ $\frac{1}{2} =$ _____ $\frac{1}{2} =$ _____

4. Use a 1 third piece. Find and record two exact coverings.

    $\frac{1}{3} =$ _____ $\frac{1}{3} =$ _____

5. Use a 1 fifth piece. Find and record one exact covering.

    $\frac{1}{5} =$ _____

6. Find and record as many exact coverings as you can. Try to find at least 20 different ones. Count $\frac{2}{3} = \frac{4}{6}$ and $\frac{4}{6} = \frac{2}{3}$ as the same exact covering.

<u>Another Name</u>

Mathematics teaching objectives:

    . Discover equivalent fractions.

Problem-solving skills pupils <u>might</u> use:

    . Use a model.

    . Guess and check.

    . Search for other solutions.

Materials needed:

    . Circle fraction pieces

Comments and suggestions:

    . Pupil accuracy in making the coverings is very important. After finding the "obvious" coverings like $\frac{1}{2} = \frac{2}{4}$, etc. encourage pupils to look for others. Provide hints for those needing help in number 6, like "$\frac{2}{3}$ can be exactly covered. Can you find two different ways?"

    . This activity will take about 20-30 minutes to complete.

Answers:

1.   2 fourths, 3 sixths, 4 eighths, 5 tenths

2.   thirds and fifths

3. $\frac{1}{2} = \frac{2}{4}$, $\frac{1}{2} = \frac{3}{6}$, $\frac{1}{2} = \frac{4}{8}$, $\frac{1}{2} = \frac{5}{10}$

4. $\frac{1}{3} = \frac{2}{6}$, $\frac{1}{3} = \frac{4}{12}$

5. $\frac{1}{5} = \frac{2}{10}$

6.   Possibilities include combinations of the following:

$$1 = \frac{2}{2}, \frac{3}{3}, \frac{4}{4}, \frac{5}{5}, \frac{6}{6}, \frac{8}{8}, \frac{10}{10}, \frac{12}{12},$$

$$\frac{1}{2} = \frac{2}{4}, \frac{3}{6}, \frac{4}{8}, \frac{5}{10}, \frac{6}{12} \qquad \frac{1}{4} = \frac{2}{8}, \frac{3}{12} \qquad \frac{3}{4} = \frac{6}{8}, \frac{9}{12}$$

$$\frac{1}{3} = \frac{2}{6}, \frac{4}{12} \qquad \frac{2}{3} = \frac{4}{6}, \frac{8}{12}$$

$$\frac{1}{5} = \frac{2}{10} \qquad \frac{2}{5} = \frac{4}{10} \qquad \frac{3}{5} = \frac{6}{10} \qquad \frac{4}{5} = \frac{8}{10}$$

$$\frac{1}{6} = \frac{2}{12} \qquad \frac{5}{6} = \frac{10}{12}$$

ADDING AND SUBTRACTING

Needed: Circle fraction pieces

1. Use the fraction pieces. Find an answer for each problem.

    a. 2 eighths plus 3 eighths is _____.

    b. $\frac{1}{5}$ added to $\frac{2}{5}$ is _____.

    c.   1 tenth   d. $\frac{4}{12}$

            2 tenths

      + 3 tenths   + $\frac{6}{12}$

2. Use the fraction pieces to find an answer for each problem.

    a. 7 tenths take away 4 tenths is _____.

    b. $\frac{9}{12}$ take away $\frac{8}{12}$ is _____.

    c.   6 eighths   d. $\frac{2}{3}$

       – 3 eighths

                            – $\frac{1}{3}$

3. Use the fraction pieces to find a fraction for each box.

    a. 2 sixths added to [＿＿＿] is 5 sixths.

    b. [＿＿＿] take away 1 fifth is 3 fifths.

    c. [＿] + $\frac{1}{4}$ = $\frac{2}{4}$ d. $\frac{7}{12}$ – [＿] = $\frac{5}{12}$

    e.   [＿]   f. $\frac{2}{3}$ g.   3 fourths

        – $\frac{8}{10}$ + $\frac{2}{3}$ – [＿]

          $\frac{9}{10}$ [＿]   0 fourths

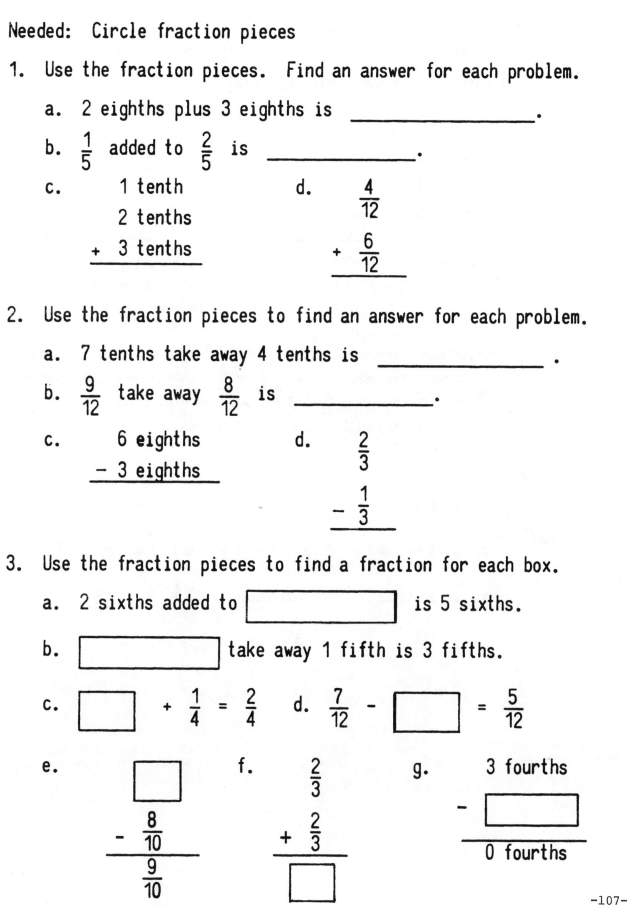

PSM 81

Adding and Subtracting

Mathematics teaching objectives:

   . Add and subtract fractions with like denominators.

Problem-solving skills pupils <u>might</u> use:

   . Use a model.

   . Look for and use patterns.

Materials needed:

   . Circle fraction pieces

Comments and suggestions:

   . Use of the fraction pieces is effective to show that 2 <u>of the</u> <u>eighths</u> <u>pieces</u> added to 3 <u>of the</u> <u>eighths</u> <u>pieces</u> equals 5 <u>of the</u> <u>eighths</u> <u>pieces</u>. Likewise use of the word names, such as 1 tenth + 2 tenths + 3 tenths = 6 tenths is effective to show that "you don't add the denominators."

   . The take-away model for subtraction works well with the fraction pieces. For example, to do $\frac{7}{12} - \frac{3}{12}$, show $\frac{7}{12}$ with the pieces and actually take away $\frac{3}{12}$. The remaining pieces show $\frac{4}{12}$.

   . This activity will take about 15 minutes to complete.

Answers:

   1. a. 5 eighths   b. $\frac{3}{5}$ c. 6 tenths   d. $\frac{10}{12}$

   2. a. 3 tenths   b. $\frac{1}{12}$ c. 3 eighths   d. $\frac{1}{3}$

   3. a. 3 sixths   b. 4 fifths   c. $\frac{1}{4}$ d. $\frac{2}{12}$

       e. $\frac{17}{10}$ f. $\frac{4}{3}$ g. 3 fourths

ADDING FRACTIONS WITH DIFFERENT DENOMINATORS

Needed: Circle fraction pieces

1. Will pieces of the first fraction _exactly_ cover the second fraction? Use the fraction pieces to decide. Write No or Yes. If yes, write how many.

   a. 1 eighth, 1 fourth _____   d. $\frac{1}{12}$, $\frac{2}{3}$ _____

   b. 1 third, 1 half _____   e. $\frac{1}{8}$, $\frac{4}{5}$ _____

   c. 1 tenth, 3 fifths _____   f. $\frac{1}{12}$, $\frac{5}{6}$ _____

2. Use the fraction pieces. Find an answer for these addition problems.

   a. 1 eighth plus 1 fourth = _____

   b. 1 tenth + 3 fifths = _____

   c. $\frac{1}{12}$ + $\frac{5}{6}$ = _____   e. $\frac{2}{6}$ + $\frac{3}{12}$ = _____

   d. $\frac{2}{5}$ + $\frac{1}{10}$ = _____   f. $\quad\frac{1}{2}$
   $\quad\quad +\frac{3}{10}$

3. Will pieces of the first fraction _exactly_ cover each of the other two fractions? Use the fraction pieces to decide. Write No or Yes. If yes, write how many for each.

   a. 1 sixth, 1 half, 1 third _____ _____

   b. 1 tenth, 2 fifths, 1 half _____ _____

   c. $\frac{1}{12}$, $\frac{2}{4}$, $\frac{1}{3}$ _____ _____   d. $\frac{1}{12}$, $\frac{2}{3}$, $\frac{3}{4}$ _____ _____

Adding Fractions With Different Denominators

Mathematics teaching objectives:

- Find equivalent fractions.
- Add fractions with different denominators.

Problem-solving skills pupils <u>might</u> use:

- Use a model.
- Solve an easier, related problem.

Materials needed:

- Circle fraction pieces

Comments and suggestions:

- Each group of problems will need teacher demonstration. Pupils are practicing finding equivalent fractions in problems 1 and 3. They will apply this method to adding in problems 2 and 4.
- Encourage a recording scheme where pupils write the "exact coverings" they have used. For example, 1 tenth + 3 fifths becomes 1 tenth + 6 tenths or 7 tenths.
- Since this packet of materials has not stressed reducing to lowest terms, pupils may find several equivalent answers as they work with the fraction pieces.

Answers:

1. a. Yes, 2   b. No   c. Yes, 6   d. Yes, 8   e. No   f. Yes, 10

2. a. 3 eighths b. 7 tenths   c. $\frac{11}{12}$ d. $\frac{5}{10}$ e. $\frac{7}{12}$ f. $\frac{8}{10}$

3. a. Yes, 3, 2   b. Yes, 4, 5   c. Yes, 6, 4   d. Yes, 8, 9

4. a. 5 sixths   b. 9 tenths   c. $\frac{7}{12}$ d. $\frac{17}{12}$

5. a. 3 eighths   b. $\frac{1}{3}$ or $\frac{4}{12}$ c. $\frac{1}{2}$

4. Use the fraction pieces. Find an answer for these addition problems.

   a.   1 half
        + 1 third
        _____

   b. 2 fifths added to 1 half = _____

   c. $\frac{1}{4}$ + $\frac{1}{3}$ = _____

   d. $\frac{2}{3}$

        + $\frac{3}{4}$
        _____

5. Use the fraction pieces to find a fraction for each box.

   a. 1 fourth plus [   ] is 7 eighths.

   b. $\frac{1}{4}$

        + [   ]
        _____
          $\frac{7}{12}$

   c. [   ] + $\frac{1}{3}$ + $\frac{1}{6}$ = 1

MORE SUBTRACTING FRACTIONS

Needed: Circle fraction pieces

1. Will pieces of the first fraction <u>exactly</u> cover the second
    fraction? Use the fraction pieces to decide. Write <u>No</u> or <u>Yes</u>.
    If yes, write how many.

    a. 1 sixth, 1 third _____   c. $\frac{1}{12}$, $\frac{1}{4}$ _____

    b. 1 eighth, 1 fourth _____   d. $\frac{1}{6}$, $\frac{1}{2}$ _____

2. Use the fraction pieces. Find an answer for these subtraction
    problems.

    a. 1 third minus 1 sixth = _____

    b. 3 fourths take away 3 eighths = _____

    c. 2 fifths subtract 1 tenth = _____   d. $\frac{5}{6} - \frac{2}{3} =$ _____

    e. $\frac{11}{12} - \frac{2}{4} =$ _____   f. $\frac{4}{6} - \frac{1}{2} =$ _____

3. Will pieces of the first fraction <u>exactly</u> cover the other two
    fractions? Use the fraction pieces to decide. Write <u>No</u> or
    <u>Yes</u>. If yes, write how many.

    a. 1 sixth;   1 half and 1 third   _____ _____
    b. 1 tenth;   1 half and 3 fifths   _____ _____
    c. 1 twelfth;   1 third and 1 fourth   _____ _____
    d. 1 twelfth;   1 third and 3 fourths   _____ _____

4. Use the fraction pieces. Find an answer for these problems.

    a. 1 half take away 1 third = _____
    b. 3 fifths take away 1 half = _____

    c. $\frac{1}{3} - \frac{1}{4} =$ _____   d. $\frac{3}{4} - \frac{1}{3} =$ _____

More Subtracting Fractions

Mathematics teaching objectives:

    . Find equivalent fractions.

    . Subtract fractions with unlike denominators.

Problem-solving skills pupils might use:

    . Use a model.

    . Solve an easier, related problem.

Materials needed:

    . Circle fraction pieces

Comments and suggestions:

    . The problems will need teacher demonstrations. Pupils are practicing finding equivalent fractions in problem 1 and 3. They will apply this method to subtracting in problems 2 and 4.

    . The take-away model for subtraction emphasizes that before 1 sixth can be taken away from 1 third, the 1 third must be traded for sixth pieces. Encourage a recording scheme as illustrated in the previous activity.

Answers:

1.   a. Yes, 2   b. Yes, 2   c. Yes, 3   d. Yes, 3

2.   a. 1 sixth   b. 3 eighths   c. 3 tenths   d. $\frac{1}{6}$ e. $\frac{5}{12}$ f. $\frac{1}{6}$

3.   a. Yes, 3, 2   b. Yes, 5, 6   c. Yes, 4, 3   d. Yes, 4, 9

4.   a. 1 sixth   b. 1 tenth   c. $\frac{1}{12}$ d. $\frac{5}{12}$

JILL AND JACK

Needed: Circle fraction pieces

Use the fraction pieces to solve these problems.

1. Jack has 3 eighths of a pizza. Jill plans to take some and leave him with 1 eighth of a pizza pie. How much will Jill take?

2. Jack and his father baked 1 whole cake. Jill and her friends ate 5 eighths of the cake. What's left for Jack and his dad?

3. One and one-fourth pies are left after dinner. For a snack Jack and Jill eat one-half of a pie. What's left for mom and dad?

4. One half of a pizza is all that's left of the $1\frac{1}{8}$ pizza pies Jack brought home. If Jill was the one that ate the pizza, how much did she eat?

5. What will Jack have left when Jill takes $1\frac{1}{2}$ of his $2\frac{2}{3}$ pies?

PSM 81

Jill And Jack

Mathematics teaching objectives:

. Solve story problems with fractions.

. Add and subtract fractions.

Problem-solving skills pupils <u>might</u> use:

. Use a model.

. Work backwards.

Materials needed:

. Circle fraction pieces

Comments and suggestions:

. If pupils work in pairs, the combined sets of circle fraction
  pieces can be used to show and do all these problems except (5).
  Encourage pupils to write a mathematical statement describing the
  problem.

. Each of these problems can be solved using the fraction pieces as
  a model. The situations could be acted out giving pupils a chance
  to talk about the processes they used.

Answers:

1. 2 eighths

2. 3 eighths

3. 3 fourths   (Some might answer one-half of a pie plus one-fourth
    of a pie. and . Practically, this answer seems more
    correct. You can show the pieces can be put together to make
    or $\frac{3}{4}$.)

4. $\frac{5}{8}$ (Watch for one-half plus one-eighth ).

5. $1 \frac{1}{6}$ or $1 \frac{2}{12}$

GROUPS AND GROUPS

Needed: Circle fraction pieces

1. a. How many groups of 2 sixths are shown? _____
    b. How many sixths in all? _____

2. a. How many groups of 1 fourth are shown? _____
    b. How many fourths in all? _____

Use the fraction pieces.

3. a. Show 2 groups of 3 eighths.
    b. How many eighths in all? _____   2 x 3 eighths = _____

4. a. Show 4 groups of 2 tenths.
    b. How many tenths in all? _____ $4 \times \frac{2}{10}$ = _____

5. a. Show 5 groups of $\frac{1}{12}$.
    b. How many twelfths in all? _____ $5 \times \frac{1}{12}$ = _____

6. a. Show 3 groups of $\frac{3}{10}$.
    b. How many tenths in all? _____ $3 \times \frac{3}{10}$ = _____

7. a. 4 groups of $\frac{1}{8}$ = _____   b. 2 groups of $\frac{2}{5}$ = _____

8. a. 3 groups of $\frac{1}{7}$ = _____   c. 4 groups of $\frac{2}{11}$ = _____

    b. 2 groups of $\frac{4}{9}$ = _____   d. 3 groups of $\frac{7}{15}$ = _____

PSM 81

Groups and Groups

Mathematics teaching objectives:

. Understand concept of multiplying fractions by whole numbers.

. Multiply fractions by whole numbers.

Problem-solving skills pupils might use:

. Use a model.

. Look for and use a pattern.

Materials needed:

. Circle fraction pieces

Comments and suggestions:

. This activity emphasizes the "groups of" approach to multiplication. That is, 3 x 2 eighths means 3 groups of 2 eighths pieces. 2 x 3 eighths means 2 groups of 3 eighths pieces. Although the answer in each case is 6 eighths, the physical representation is different.

. The use of word names helps eliminate the tendency of saying that 3 groups of $\frac{2}{8}$ is $\frac{6}{24}$.

. If you have transparent circle pieces, the above concept can be shown by placing the twelfths pieces on the overhead and asking pupils to show 3 groups of 4 twelfths, 4 groups of 3 twelfths, etc.

. This activity will take about 20 minutes to complete.

Answers:

 1. a. 2   b. 4 sixths

 2. a. 3   b. 3 fourths

 3. b. 6, 6 eighths

 4. b. 8, 8 tenths

 5. b. 5, $\frac{5}{12}$

 6. b. 9, $\frac{9}{10}$

 7. a. $\frac{4}{8}$ b. $\frac{4}{5}$

 8. a. $\frac{3}{7}$ b. $\frac{8}{9}$ c. $\frac{8}{11}$ d. $\frac{21}{15}$

MORE GROUPS

Needed: Circle fraction pieces

1. a. Make a group of 2 eighths.

    b. Make $\frac{1}{2}$ of a group of 2 eighths.

    c. What is $\frac{1}{2}$ of a group of 2 eighths? _____

2. a. Make a group of $\frac{8}{12}$.

    b. Make $\frac{1}{4}$ of a group of $\frac{8}{12}$.

    c. What is $\frac{1}{4}$ of $\frac{8}{12}$? _____

3. a. Make a group of $\frac{8}{10}$.

    b. Make $\frac{3}{4}$ of a group of $\frac{8}{10}$.

    c. $\frac{3}{4}$ of $\frac{8}{10}$ = _____

4. Use the fraction pieces. Find an answer for these multiplication patterns.

    a. $\frac{1}{3}$ of a group of 3 sixths = _____

    b. $\frac{1}{5}$ of a group of $\frac{10}{10}$ = _____

    c. $\frac{2}{3}$ of a group of 9 twelfths = _____

    d. $\frac{3}{4}$ of a group of $\frac{4}{12}$ = _____

    e. $\frac{1}{4}$ of $\frac{8}{10}$ = _____   f. $\frac{3}{6}$ of $\frac{12}{12}$ = _____

5. Use the fraction pieces. Find an answer that uses the fewest number of pieces.

    a. $\frac{1}{2}$ of $\frac{6}{12}$ = _____   c. $\frac{1}{3}$ of $\frac{6}{8}$ = _____

    b. $\frac{1}{2}$ of $\frac{10}{10}$ = _____   d. $\frac{2}{4}$ of $\frac{8}{10}$ = _____

More Groups

Mathematics teaching objectives:

. Understand concept of multiplication of fractions. (Readiness)

. Multiply fraction by another fraction. (Readiness)

Problem-solving skills pupils might use:

. Use a model.

Materials needed:

. Circle fraction pieces

Comments and suggestions:

. Pupils need this understanding of fractions to do this activity.

> Two-thirds of a group means to separate the group into three equal parts and then use two of the equal groups. For example, $\frac{2}{3}$ of a group of $\frac{6}{8}$ means to separate the $\frac{6}{8}$ into three equal groups of $\frac{2}{8}$ each and then use two of the groups to get $\frac{4}{8}$.

Work with transparent fraction pieces on the overhead can introduce and reinforce this concept.

. Problem 5 requires pupils to do some "exact coverings" to find the fewest number of pieces. This, of course, is a physical way of illustrating reducing fractions.

. Note: This activity, except in special cases like $\frac{1}{2}$ x $\frac{1}{3}$ (or $\frac{1}{2}$ x $\frac{2}{6}$) = $\frac{1}{6}$, does not lead to the algorithm of multiply numerators and multiply denominators. But a definite advantage of using the process shown in this activity is that pupils can find the answer without knowing the algorithm.

Answers:

1. c. 1 eighth   2. c. $\frac{2}{12}$ 3. c. $\frac{6}{10}$

4. a. 1 sixth   b. $\frac{2}{10}$ c. 6 twelfths   d. $\frac{1}{12}$ e. $\frac{2}{10}$ f. $\frac{6}{12}$

5. a. $\frac{1}{4}$ b. $\frac{1}{2}$ c. $\frac{1}{4}$ d. $\frac{2}{5}$

STILL MORE GROUPS

Needed: Circle fraction pieces

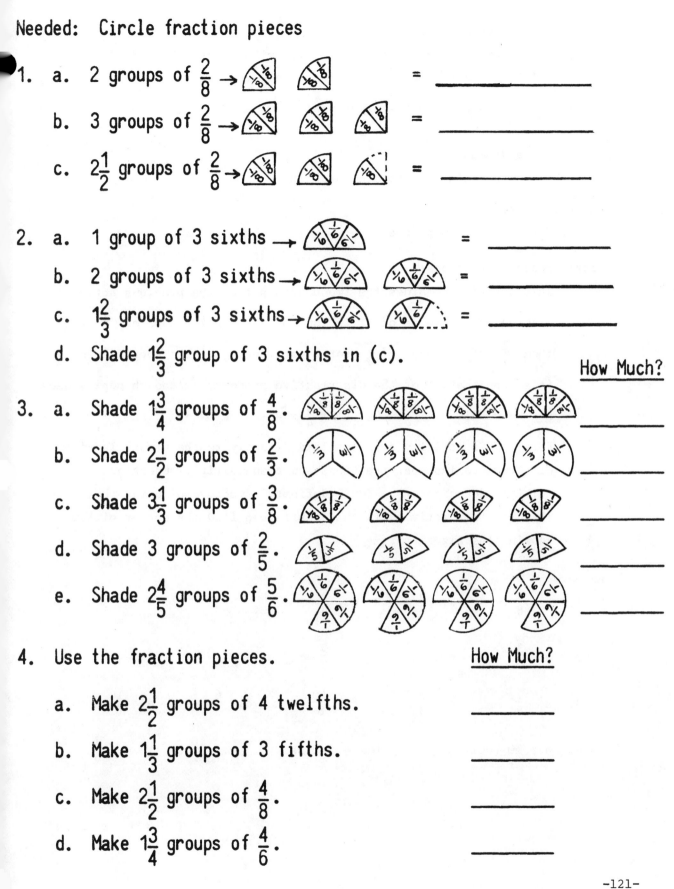

1. a. 2 groups of $\frac{2}{8}$ → = _____

    b. 3 groups of $\frac{2}{8}$ → = _____

    c. $2\frac{1}{2}$ groups of $\frac{2}{8}$ → = _____

2. a. 1 group of 3 sixths → = _____

    b. 2 groups of 3 sixths → = _____

    c. $1\frac{2}{3}$ groups of 3 sixths → = _____

    d. Shade $1\frac{2}{3}$ group of 3 sixths in (c).

How Much?

3. a. Shade $1\frac{3}{4}$ groups of $\frac{4}{8}$. _____

    b. Shade $2\frac{1}{2}$ groups of $\frac{2}{3}$. _____

    c. Shade $3\frac{1}{3}$ groups of $\frac{3}{8}$. _____

    d. Shade 3 groups of $\frac{2}{5}$. _____

    e. Shade $2\frac{4}{5}$ groups of $\frac{5}{6}$. _____

4. Use the fraction pieces.   How Much?

    a. Make $2\frac{1}{2}$ groups of 4 twelfths.   _____

    b. Make $1\frac{1}{3}$ groups of 3 fifths.   _____

    c. Make $2\frac{1}{2}$ groups of $\frac{4}{8}$. _____

    d. Make $1\frac{3}{4}$ groups of $\frac{4}{6}$. _____

-121-

Mathematics teaching objectives:

. Understand concept of multiplication with fractions. (Readiness)

. Multiply fraction by a mixed fraction. (Readiness)

Problem-solving skills pupils might use:

. Use a model.

. Use a drawing.

. Solve an easier but related problem.

Materials needed:

. Circle fraction pieces

Comments and suggestions:

. This activity combines the two concepts from the previous activities. Finding $2\frac{1}{2}$ groups of $\frac{2}{8}$ is done by finding 2 groups of $\frac{2}{8}$ and then $\frac{1}{2}$ of a group of $\frac{2}{8}$ for a total of $\frac{5}{8}$. The approach actually is an application of the distributive property, although pupils need not know this. $1\frac{3}{4} \times 4 = (1 + \frac{3}{4}) \times 4 = (1 \times 4) + (\frac{3}{4} \times 4) = 4 + 3 = 7$

. Note: This activity does not lead directly to the usual algorithm of change to improper fractions and then multiply numerators and multiply denominators. But a definite advantage of using the process shown in this activity is that pupils can find the answer without knowing the algorithm.

Answers:

1. a. $\frac{4}{8}$ b. $\frac{6}{8}$ c. $\frac{5}{8}$

2. a. $\frac{3}{6}$ b. $\frac{6}{6}$ c. $\frac{5}{6}$

3. a. $\frac{7}{8}$ b. $\frac{5}{3}$ c. $\frac{10}{8}$ d. $\frac{6}{5}$ e. $\frac{14}{6}$

4. a. 10 twelfths   b. 4 fifths   c. $\frac{10}{8}$ d. $\frac{7}{6}$

EQUAL SHARES

Needed: Circle fraction pieces

1. Sam has a group of 9 twelfths. He wants to separate them into 3 equal groups. How big will each group be?

2. Carla has a group of $\frac{10}{12}$. How many groups of $\frac{2}{12}$ can she make?

3. How big was the group if Lonnie was able to get 4 groups of $\frac{2}{10}$ from it?

4. Arthur has a group of 6 one-eighth pieces. He wants to separate them into smaller groups all the same size. How many different ways can he do it? Write the number of groups and the size of each smaller group.

5. How many different ways can 8 twelfths be divided up into smaller groups? Write the number of groups and the size of each smaller group.

6. Maria has less than 2 whole circles. She divided the fraction pieces into equal groups of 3 eighths. How many eighths could she have started with?

7. This diagram shows two division problems.

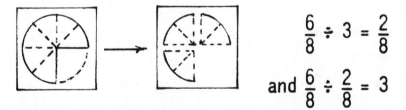

$$\frac{6}{8} \div 3 = \frac{2}{8}$$

and $\frac{6}{8} \div \frac{2}{8} = 3$

Write two division problems for this diagram.

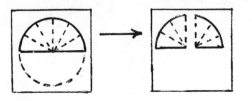

Equal Shares

Mathematics teaching objectives:

. Understand concept of division with fractions.

. Divide fraction by another fraction or by a whole number.

Problem-solving skills pupils might use:

. Use a model.

. Search for and be aware of other solutions.

Materials needed:

. Circle fraction pieces

Comments and suggestions:

. The operation of division is the hardest for pupils to do. The concept of division, using the circle fraction pieces, is much easier. Division problems ask either how many groups there are or how large is each group. For example, $\frac{6}{8} \div 3$ means how big is each group if $\frac{6}{8}$ is divided into 3 equal groups. And $\frac{6}{8} \div \frac{2}{8}$ means how many groups are there if $\frac{6}{8}$ is divided into equal groups of $\frac{2}{8}$ each.

. To emphasize both of the above meanings, this activity might work best as a teacher directed activity.

. Note: This activity does not lead to the usual algorithm for dividing fractions. But a definite advantage of the process shown in this activity is that pupils can get answers to certain division situations without knowing the algorithm.

Answers:

1. 3 twelfths

2. 5 groups

3. $\frac{8}{10}$

4. 6 groups of $\frac{1}{8}$, 3 groups of $\frac{2}{8}$, 2 groups of $\frac{3}{8}$

5. 8 groups of $\frac{1}{12}$, 4 groups of $\frac{2}{12}$, 2 groups of $\frac{4}{12}$

6. $\frac{3}{8}$, $\frac{6}{8}$, $\frac{9}{8}$, $\frac{12}{8}$, $\frac{15}{8}$

7. $\frac{6}{12} \div 2 = \frac{3}{12}$ or $\frac{6}{12} \div \frac{3}{12} = 2$

FRACTION PATTERNS

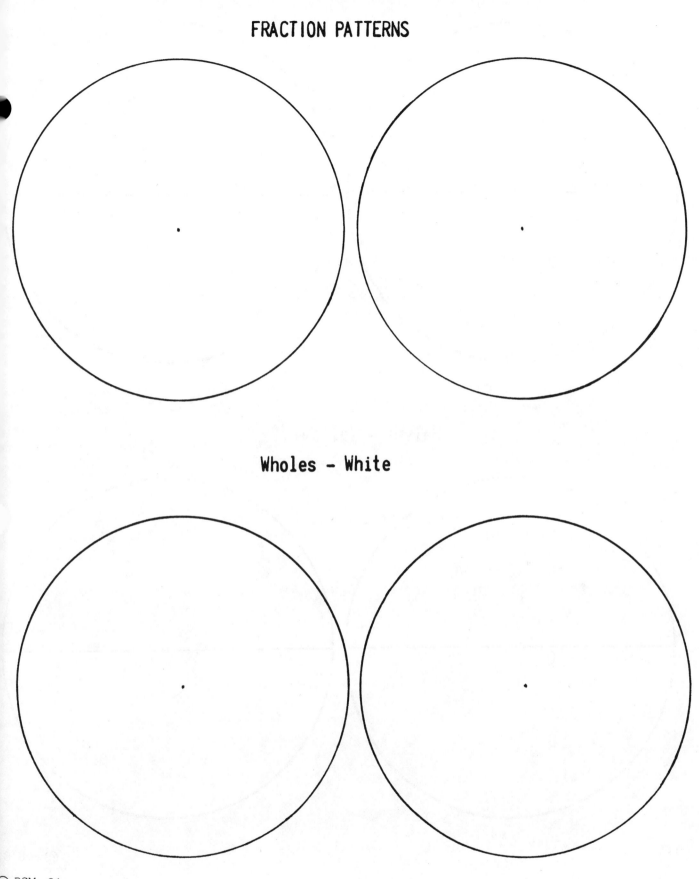

Wholes - White

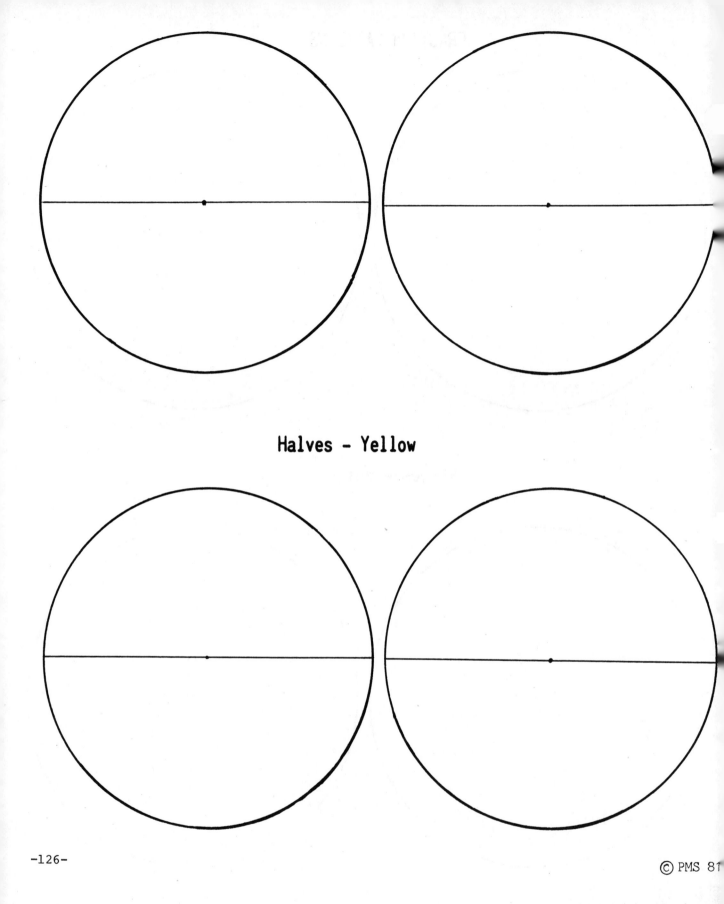

Halves - Yellow

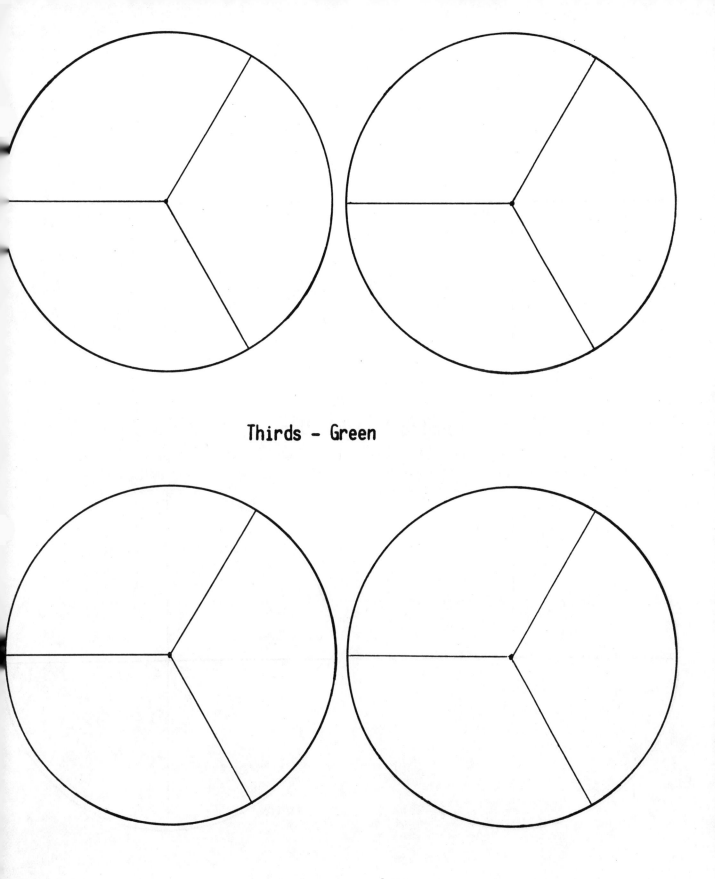

Thirds - Green

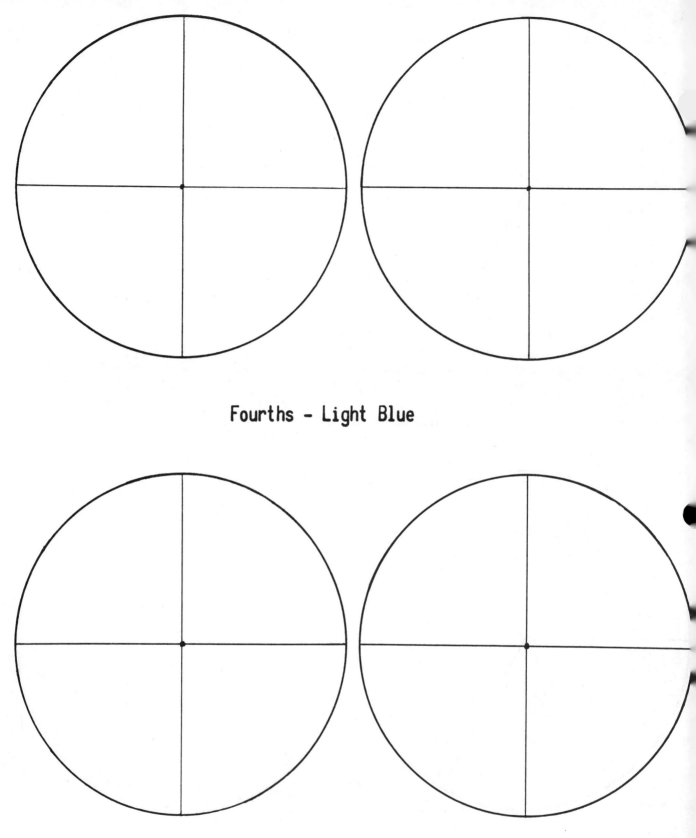

Fourths - Light Blue

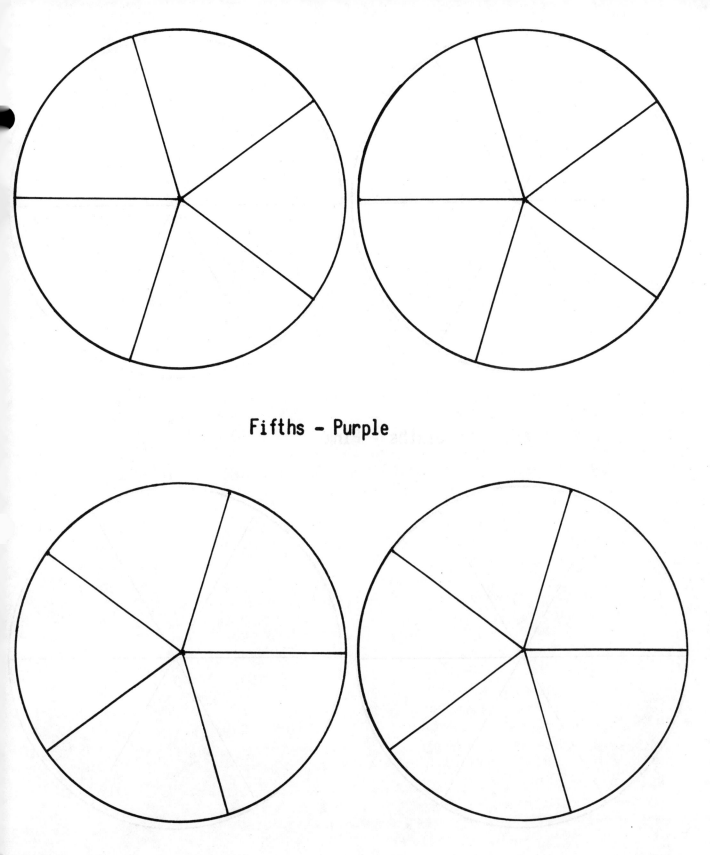

Fifths - Purple

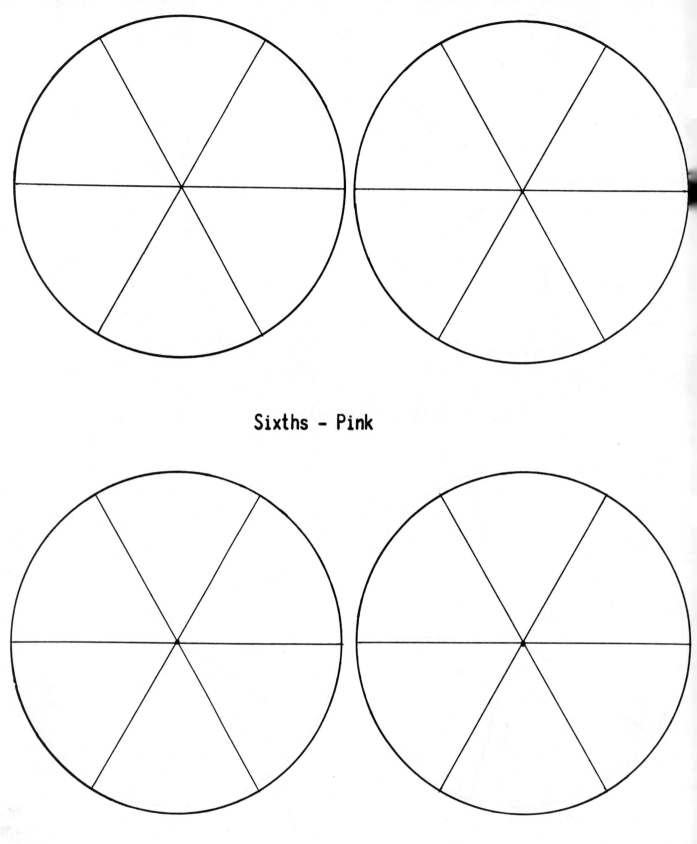

Sixths - Pink

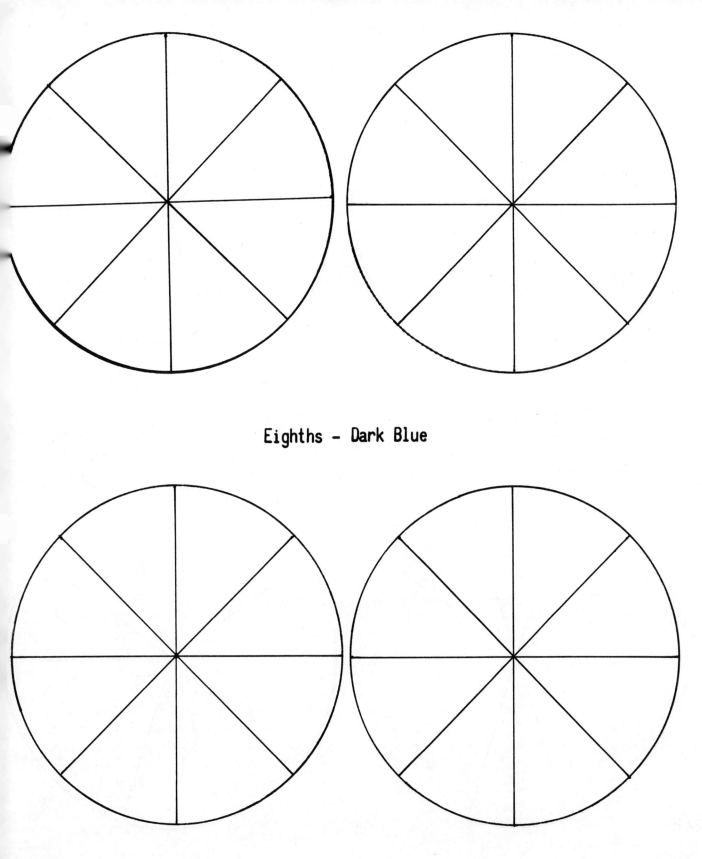

Eighths - Dark Blue

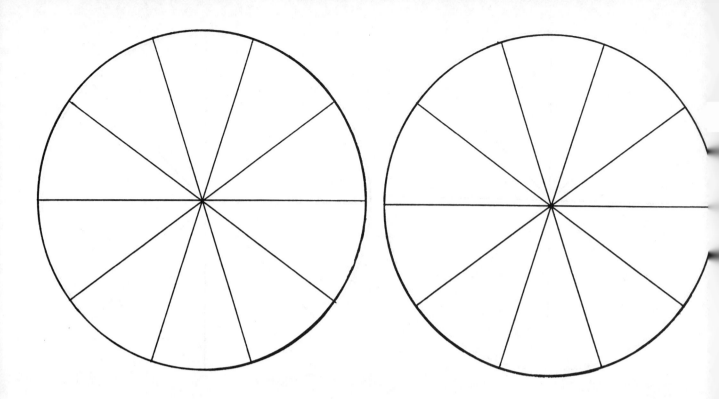

Tenths - Red

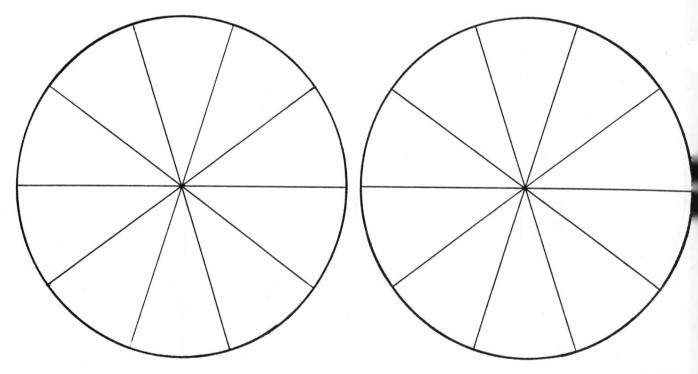

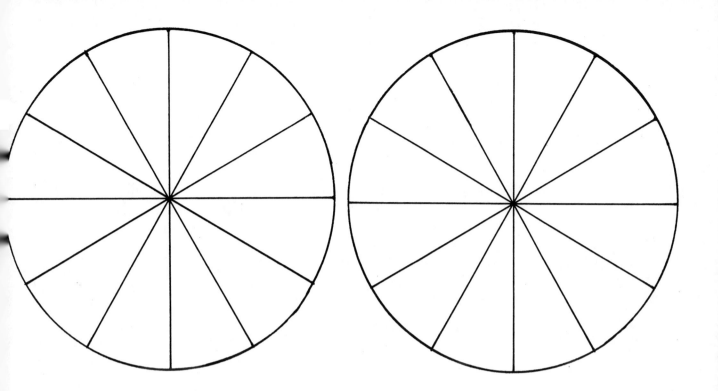

Twelfths - Orange

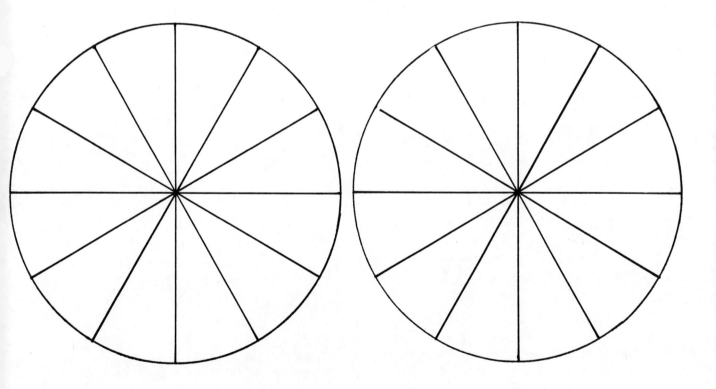

Grade 5

V. GEOMETRY

V. GEOMETRY

Children have a natural interest in geometric ideas because they live in a world filled with shapes and solids. The middle school years provide an excellent opportunity to continue the study of space relationships and the attributes of geometric objects. Fortunately this study can also be used to improve the ability of students to visualize, to allow the "number poor" but "geometric rich" student to shine and to emphasize problem solving skills.

The 3-year old structural engineer . . .

. . . might become your geometry star!

I might not get long division but I'm a whiz at recognizing shapes and patterns!

There is little agreement on grade placement for specific geometric topics, however the activities included here relate to materials included in many fifth-grade texts. The activities fall into four main categories: (1) lines and squares on a geoboard, (2) attributes of 3-dimensional solids-- cubes, pyramids, spheres, cylinders, etc. (3) volume concepts and (4) symmetry. Besides emphasizing problem solving, the activities incorporate the use of physical models and their drawings, verbal descriptions and names. An emphasis is placed on visualization because many students have difficulties with geometry, drafting and general problem solving due to poorly developed ability to visualize.

Using the Activities

The activities can be used when related topics are covered in the textbook or they may be used at any time during the year. The required arithmetic skills are simple with some multiplication used in "Solid Boxes." Possible problems students might have with the geometry are discussed on each page. A set of geometric models is needed for several of the activities. These models can be purchased commercially, made from tagboard or collected from everyday objects. The page, "Solid Stations," has more suggestions on models.

Many of the activities are meant for small groups or partners. An activity can be done by one group while the rest are doing other work at their desks. Some activities might be used at interest stations for pupils with unassigned time. If you have enough material, several groups can do the activity at one time.

LINE SEGMENTS

Geoboard, rubber bands, metric ruler.

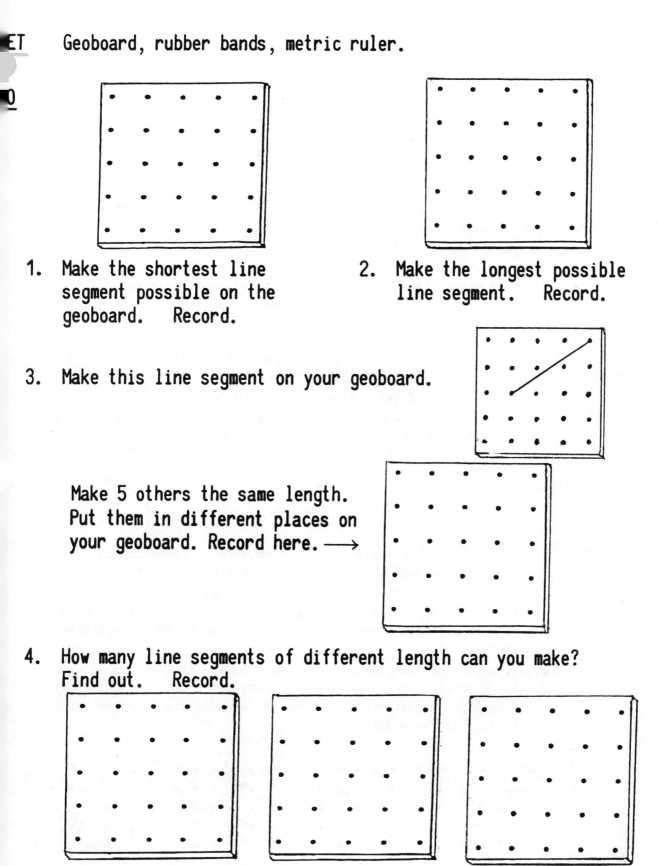

1. Make the shortest line segment possible on the geoboard.   Record.

2. Make the longest possible line segment.   Record.

3. Make this line segment on your geoboard.

Make 5 others the same length. Put them in different places on your geoboard. Record here. ⟶

4. How many line segments of different length can you make? Find out.   Record.

Line Segments

Mathematics teaching objectives:

. Measure and compare lengths of line segments using a metric ruler.
. Measure line segments to the nearest millimetre.
. Visualize shapes in different positions.

Problem-solving skills pupils _might_ use:

. Make and use a drawing or physical model.
. Search for and be aware of other possibilities.
. Record answers systematically.

Materials needed:

. A classroom set of geoboards and rubber bands.
. Supply of geoboard record paper.
. Metric rulers labeled in centimetres and subdivided into millimetres.
. Transparent geoboard for use on overhead projector. (Optional)

Comments and suggestions:

. Let pupils work the page on their own. Then take some time for pupils to compare their solutions.

. Pupils might need help using a metric ruler.

. During a discussion near the end of the activity, emphasize a systematic way for working and recording problem 4. (The creativity of some pupils may be inhibited by imposing a system on them.

Answers:

1. ●—● or ▮   2. ⟍⟍⟍ or ⟋⟋⟋

3. 23 other line segments can be drawn with the same length as the given segment. Each segment is the diagonal of a 2 by 3 or 3 by 2 rectangle. (During discussions, introduce and use the term diagonal of a rectangle.) If pupils want to find as many solutions as possible, encourage them to make a systematic search.

4. 14 different lengths are possible. Notice the systematic way of recording. The last drawing shows how all 14 lengths can be drawn from a corner nail connecting each of the 14 other nails in one-half of a geoboard.

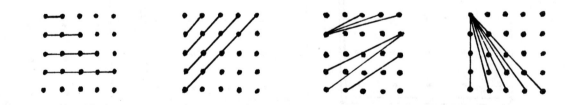

JUST FOR SQUARES

<u>GET</u>   Geoboard, rubberbands, metric ruler, index card

<u>DO</u>

1. Make a square on your geoboard.
    Record below.

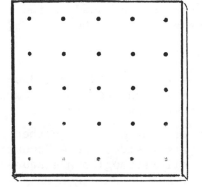

Tell what is special about the sides of a square. (Hint: use the ruler.)

_____

_____

_____

Tell what is special about the corners of a square. (Hint: use the index card.)

_____

_____

_____

2. Make as many different-sized squares as you can.
    Try to find at least 6. Record each.

Mathematics teaching objectives:

. Make different-sized squares on a geoboard.
. Identify properties of a square.

Problem-solving skills pupils might use:

. Make and/or use drawings or physical models.
. Record solution possibilities.
. Look at a problem from various points of view.

Materials needed:

. Classroom set of 25-point geoboards and rubber bands of various colors.
. A transparency of the 25-point pattern can be used on the overhead. Shapes can be drawn with colored pens made for the overhead.

Comments and suggestions;

. The first problem can be solved by a teacher-led class discussion. Pupils can work the rest of the activity on their own. Save some time near the end of the period for sharing results.

. Pupils might think squares are not rectangles. In another discussion, you might explain that a rectangle has 4 sides and 4 "square" angles. All squares are rectangles but not all rectangles are squares.

Answers:

1. There are 4 sides. They all have the same length. A square has 4 square corners or 4 equal angles.

2. There are eight different-sized squares. The last four are harder to find. Pupils often express delight when they first find a tilted square.

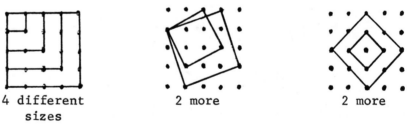

4 different   2 more   2 more
sizes

Challenge: Answers will vary. Pupils will enjoy showing their designs as they are made. A bulletin board display could be made of the various designs pupils have created. Two examples are shown.

Just For Squares (cont.)

2. (cont.)

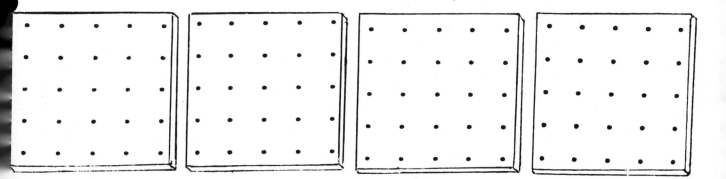

<u>CHALLENGE</u>   Make some geoboard designs that use only squares.
Record below.

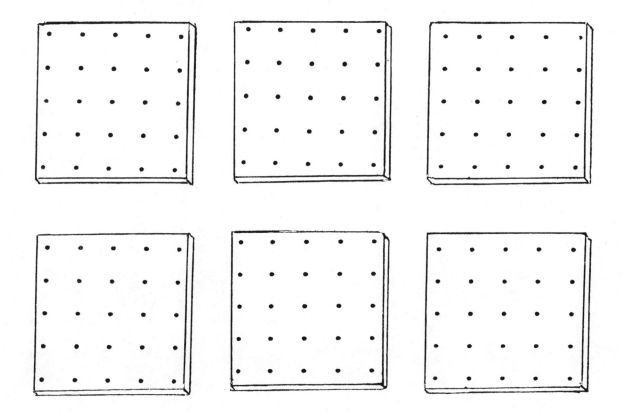

PSM 81

SOLID STATIONS

The three activities that follow use geometric models.

Get a set of wooden, plastic, or cardboard models including the following: cubical blocks, rectangular blocks, spheres, cylinders, pyramids, cones.

If commercial sets of geometric models are not available, patterns are included to make a cube, a rectangular block, a triangular block, a pyramid, and a cylinder. Sturdy models can be made by reproducing these patterns on tagboard. Pupils can also assist in collecting "environmental substitutes," such as tin cans, blocks, boxes, paper cups, balls, wedges, etc.

1. Label each model: A, B, C, etc.

2. Place an activity card and a set(s) of models in an interest area.

3. Allow pupils some "free play" time.

4. Have pupils complete the activity.

5. Plan time to discuss and compare results.

Cone

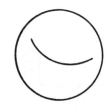

Sphere

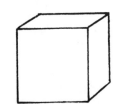

Cube

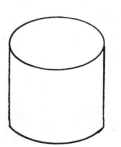

Cylinder

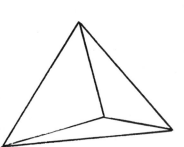

Tetrahedron

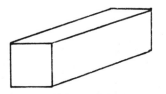

Rectangular block

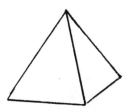

Pyramid

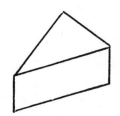

Triangular block

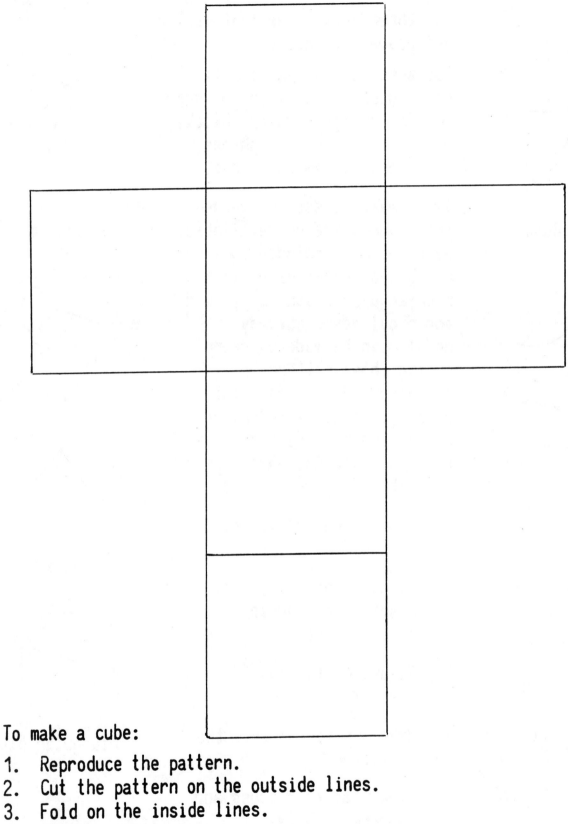

To make a cube:

1. Reproduce the pattern.
2. Cut the pattern on the outside lines.
3. Fold on the inside lines.
4. Secure with tape.

Solid Stations (cont.)

To make a rectangular block:

1. Reproduce the pattern.
2. Cut the pattern on the outside lines.
3. Fold on the inside lines.
4. Secure with tape.

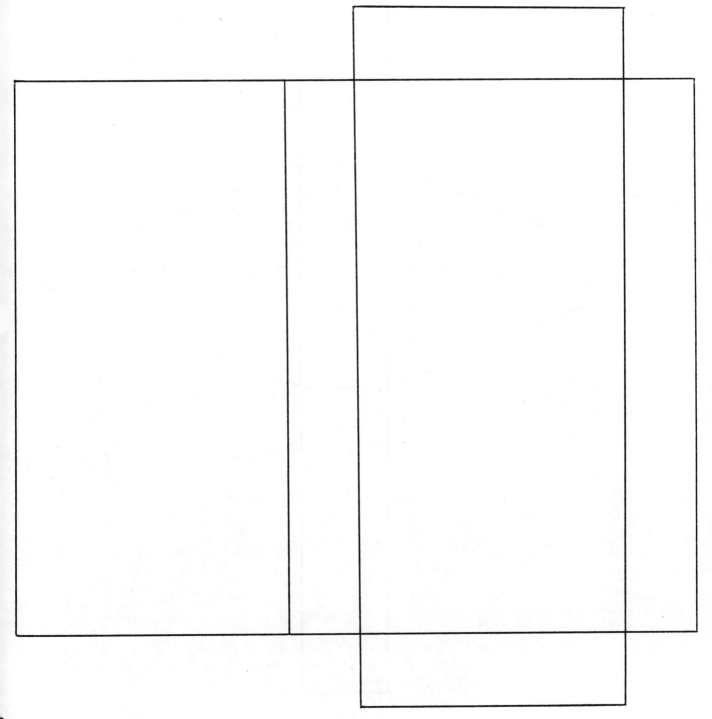

Solid Stations (cont.)

To make a triangular block:

1. Reproduce the pattern.
2. Cut out the pattern on the outside lines.
3. Fold on the inside lines.
4. Secure with tape.

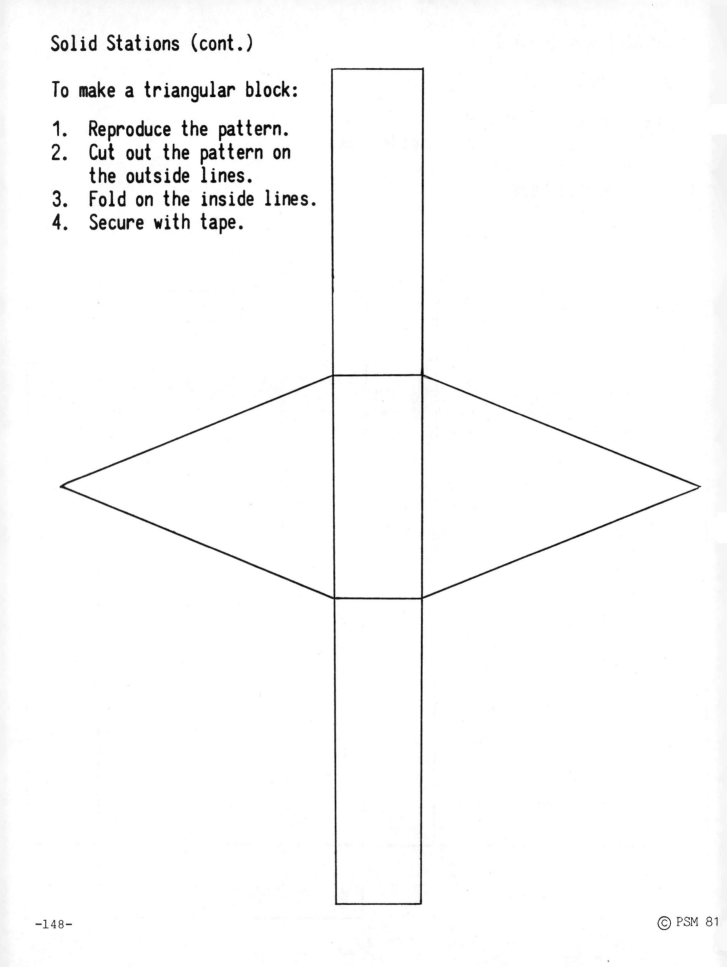

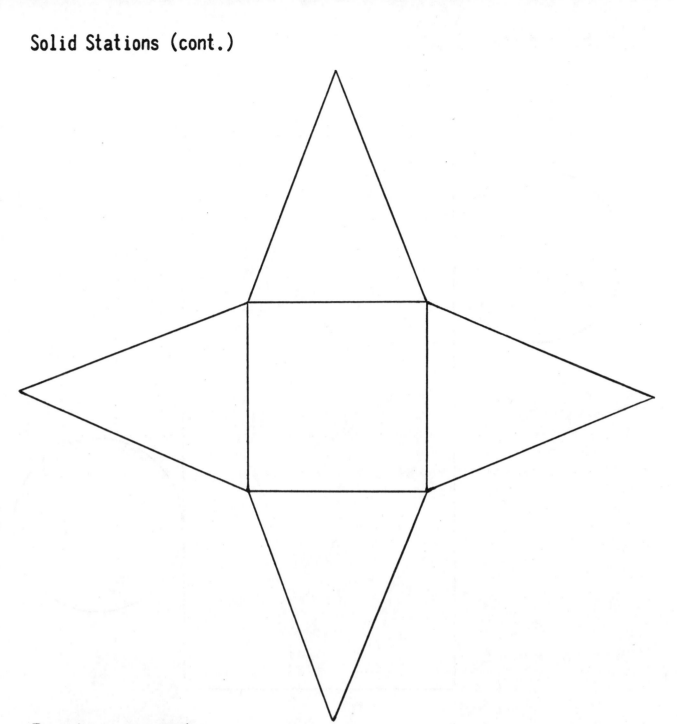

To make a pyramid:

1. Reproduce the pattern.
2. Cut out the pattern on the outside lines.
3. Fold on the inside lines.
4. Secure with tape.

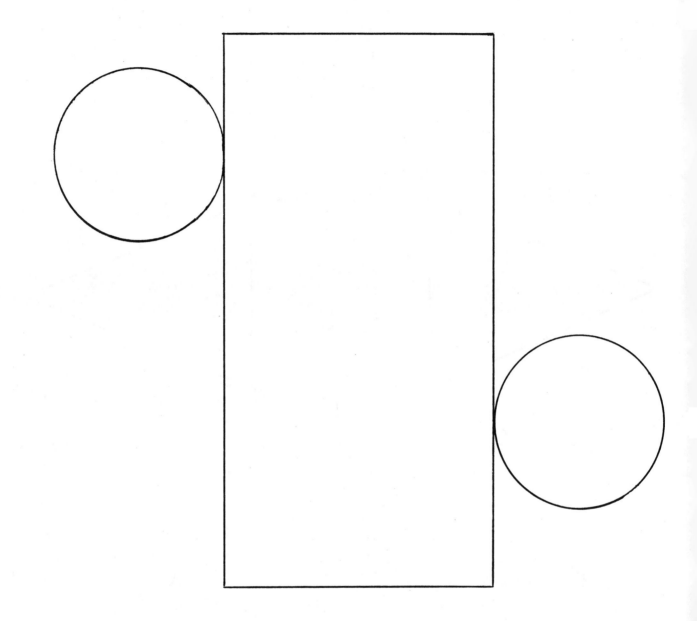

To make a cylinder:

1. Reproduce the pattern.
2. Cut out the pattern. Don't cut off the circles.
3. Shape into a cylinder.
4. Secure with tape.

ROLL ON

1. Find all the objects that roll easily.
   List them. _____

2. Which objects roll in a straight line?
   List. _____

3. Which objects roll but not in a straight line?
   List. _____

4. Some of the objects in questions 1, 2, and 3 can be placed
   so they won't roll.
   List them. _____

5. Which objects do not roll?
   List. _____

6. What is the property of an object that allows it to roll?

   _____

7. List as many things in the classroom as you can that will
   roll. List the geometric model that each thing is most alike.

   _____

   _____

   _____

Roll On

Mathematics teaching objectives:

. Recognize attributes of 3-dimensional objects.

. Develop 3-dimensional concepts.

Problem solving skills pupils _might_ use:

. Recognize attributes of an object.

. Classify objects.

. Use a physical model.

. Guess and check.

Materials needed:

. Geometric solids or suitable everyday objects. See page 145, "Solid Stations," for ideas.

Comments and suggestions:

. Have pupils bring examples of geometric solids from home. You may be able to collect enough materials for several sets.

. Because of the materials needed, this activity is best used with small groups of pupils. Perhaps part of the class could work on this and the next two activities while the remainder of the class works at their desks or at other stations.

. Names of the solids can be introduced by labeling each object as a cone, cube, ...   Vocabulary can be avoided by labeling all cones with A, cubes with B, ...   A student could help with the labeling by looking at the pupil page Solid Stations.

Answers:

Answers will vary according to the objects used. Pupils should recognize that having a curved surface allows a solid to roll.

STACKS

1. Find all the objects that stack easily.
    List them. _____

2. Be sure your partner has the same objects as you have.

3. Build a stack of objects so your partner cannot see it.

4. Describe your stack very carefully. Your partner will
    build exactly what you tell him to build.

5. Look at both stacks to see if they are the same.

6. Make a sketch of your stack. Do a front sketch, a side
    sketch, and a top sketch.

7. See if another classmate can build your stack by looking
    at your sketch.

Stacks

Mathematics teaching objectives:

. Develop 3-dimensional concepts.

Problem solving skills pupils might use:

. Visualize an object from its description.

. Describe an object verbally and with diagrams.

. Use a physical model.

. Break a problem into manageable parts.

Materials needed:

. Geometric solids or suitable everyday objects.

Comments and suggestions:

. Pupils can be encouraged to describe the stacks and sketches using
geometric terms rather than common names like milk carton, tin can, etc.
A technique to use is to describe the one part at a time.

. Pupils might find it hard to sketch their stacks as required in #6.
They will have better luck if they view the stack with one eye closed
directly from the top, front or side.

. An alternative to having one student build a stack and then describe
it to a partner is to have a third person describe a stack to two pupils.
The pupils could then compare completed stacks and note differences.
Cooperation skills are promoted using this approach.

. Because of the materials needed this activity is best used with a few
pairs of pupils. Perhaps it could be one laboratory activity of several
which emphasize many different mathematical concepts. For other suggestions, see the overview to this section.

Answers:

Answers will vary.

MYSTERY SEARCH

Find an object with
Which object?

_____

· 6 faces
· 8 vertices
· 12 edges
· No square faces

Find an object with
Which object?

_____

· 1 curved surface
· 2 flat surfaces
· 2 curved edges
· a height less than the distance around

Make up two searches of your own. Use as many clues as you want. Record them below. See if a classmate can find your objects.

Find an object with

Which object?

_____

Find an object with
Which object?

_____

Mystery Search

Mathematics teaching objectives:

. Describe attributes of 3-dimensional objects.

. Develop 3-dimensional concepts.

Problem solving skills pupils might use:

. Visualize an object from its description.

. Guess and check.

. Break a problem into manageable parts.

. Invent new problems by varying one that has been solved.

Materials needed:

. Geometric objects or suitable everyday objects. See page 145, "Solid Stations," for ideas.

Comments and suggestions:

. This activity could be used with small groups or could be teacher-directed with the teacher giving the clues for the first two objects. The complete set should be displayed for pupils to look at. Each pupil then can determine the object being described.

. The vocabulary may be altered to fit the abilities of your pupils, i.e., sides instead of faces, corners instead of vertices, etc.

. When creating their own searches, pupils should check each other, making sure the clues describe only one object. Pupil-created searches could be dittoed to extend the activity.

Answers:

The first search is a rectangular prism or box with no square faces. The second search is a cylinder or tin can. The last clue implies the cylinder is not a tall, skinny one.

PYRAMIDS

Pyramids have a point, faces that are triangles and bottoms that can have different shapes. A pyramid gets its name from the shape of the bottom.

Triangular

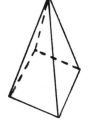
Square

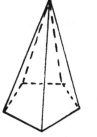
Pentagonal

Hexagonal

1. Use the drawings to help you fill in the table.

Pyramid Name	Number of sides in the bottom	Number of faces including the bottom	Number of corners	Number of edges
Triangular				6
Square		5		
Pentagonal	5			
Hexagonal			7	

2. Answer these by looking for patterns in the table.

   If the bottom of a pyramid has

   a. 8 sides, it has _____ faces.

   b. 13 sides, it has _____ corners.

   c. 50 sides, it has _____ edges.

3. Look for other patterns in the table. Write them on the back of this paper.

Pyramids

Mathematics teaching objectives:
- . Find and use number patterns.
- . Relate pictures of solids to models.

Problem solving skills pupils <u>might</u> use:
- . Use a drawing.
- . Make a systematic listing.
- . Observe patterns and then make a prediction.

Materials needed:
- . A model for each of the pyramids is helpful.

Comments and suggestions:
- . This activity can be done as a whole group. Each pupil will need an activity sheet. The vocabulary (sides, faces, edges, corners) can be explained as the models are shown.
- . Some pupils will have trouble interpreting the drawings, especially the hidden faces and edges. Actual models will help. Students can match the models to the drawings before starting on the activity. If students have made the models themselves, they will probably have a better understanding of the solids and the drawings.
- . A similar activity could be written using prisms--triangular, rectangular, pentagonal, etc. A pattern will occur in the table although it will be different than the pattern for the pyramids.

Answers:

1.

Triangular	3	4	4	6
Square	4	5	5	8
Pentagonal	5	6	6	10
Hexagonal	6	7	7	12

2. a. 9
   b. 14
   c. 100

3. Several patterns will be seen by pupils. Encourage open discussion of whatever patterns they see. One difficult pattern to see is the number of corners + number of faces = number of edges + 2. This pattern works for most 3-dimensional objects with "flat" faces.

SOLID BOXES

Get a box of cubes.

1. Build a solid box that looks
   like this.

   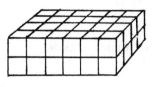

   The box used _____ cubes.

   It is _____ cubes long.

   _____ cubes wide.

   _____ cubes tall.

2. Use 48 cubes. Build a solid
   box different from the one
   in the picture.

   My box is _____ cubes long.

   _____ cubes wide.

   _____ cubes tall.

3. Build two more solid boxes. Use 48 cubes for each of them.
   Make them different. Record.

   _____ cubes long   _____ cubes long

   _____ cubes wide   _____ cubes wide

   _____ cubes tall   _____ cubes tall

4. Find some boxes you could build that use 100 cubes each.
   Make each one different.

   _____ cubes long   _____ cubes long   _____ cubes long

   _____ cubes wide   _____ cubes wide   _____ cubes wide

   _____ cubes tall   _____ cubes tall   _____ cubes tall

Solid Boxes

Mathematics teaching objectives:

    . Develop volume concepts for a rectangular prism.

    . Practice multiplication facts.

Problem solving skills pupils <u>might</u> use:

    . Use a drawing.

    . Observe patterns and then make predictions.

    . Guess and check.

    . Search for other solutions.

Materials needed:

    . Cubes--at least 48

Comments and suggestions:

    . A "different" box is one which has at least two dimensions different. A <u>6 by 4 by 2</u> is the same as a <u>4 by 6 by 2</u> but different than a <u>3 by 8 by 2</u>.

    . Some pupils may discover a "rule" for finding the volume of a rectangular prism. Use of the standard formula $V = \ell \times w \times h$ should be delayed until pupils have had several readiness activities such as this one.

    . Because of the number of cubes needed, this activity is best done with a small group. Perhaps it can be integrated with several other activities.

Answers:

1. 48 cubes--6 cubes long, 4 cubes wide, 2 cubes tall.

2.-3. The product of the dimensions should be 48. Possibilities are 1, 1, 48; 1, 2, 24; 1, 3, 16; 1, 4, 12; 1, 6, 8; 2, 2, 12; 2, 3, 8; 2, 4, 6; 3, 4, 4.

4. The product of the dimensions should be 100. Possibilities are 1, 1, 100; 1, 2, 50; 1, 4, 25; 1, 5, 20; 1, 10, 10; 2, 2, 25; 2, 5, 10; 4, 5, 5.

ALPHABET FLIPPING

1. Some letters of the alphabet can be flipped over to fit
   exactly on top of themselves. Use the letters below for a model.

   # A B C D E F G H I J K L M N
   # O P Q R S T U V W X Y Z

   a. Which letters fit with a left to right flip? _____
   b. Which letters fit with a top to bottom flip? _____
   c. Which letters will not fit with either flip? _____
   d. Which letters fit with both flips--top to
      bottom and left to right?   _____

   The line the letters are flipped over is called a <u>line of symmetry</u>.

2. Flip these half-letters over the line of symmetry to answer
   the riddles.

   a. What did Sue Ann see in her
      alphabet soup?

   b. What does Farmer
      McDonald say to
      his animals?

3. Which letters fit if you <u>turn</u> the letter top to bottom?

4. Which lower case, printed letters have a line of symmetry?

5. Make up your own line of symmetry message.

Alphabet Flipping

Mathematics teaching objectives:
- Find lines of symmetry.
- Identify objects which have symmetry.

Problem solving skills pupils **might** use:
- Recognize attributes of an object.
- Classify objects.
- Use a drawing.
- Visualize and draw an object.

Materials needed:
- None

Comments and suggestions:
- The activity can be done with the whole group. Large letters drawn on a transparency can be used to demonstrate what is meant by a top to bottom flip or a left to right flip.
- Pupils could also decide if a letter has symmetry by using a mirror, a shiny piece of metal or a piece of plastic. A letter has symmetry if it looks correct when a mirror is placed on a suspected line of symmetry.
- Holding the sheet on a window or up to a bright light and then flipping the sheet over allows pupils to identify letters that look the same.
- Pupils will have to decide how the lower case letters should be printed. Different styles will produce different answers.

Answers:

1. a. A, H, I, M, O, T, V, W, X, Y

   b. B, C, D, E, H, I, O, X

   c. F, G, J, K, L, N, P, Q, R, S, U, Z

   d. H, I, O, X

2. a. WOW, MOM HIT TOM WITH A TOMATO
   b. EIEIO

3. N, S, Z

4. C, I, L, O, T, V, W, X

TRADEMARKS

Trademarks are symbols and shapes used to identify businesses and products.

GET: An old telephone book or a newspaper

1. Find and cut out 10 trademarks from the yellow pages or the newspaper. Record the company names below. If possible, fold each trademark to make one part exactly fit on the other part. The fold line is called a <u>line</u> <u>of</u> <u>symmetry</u>.

2. Some trademarks may have more than one line of symmetry. Fold the 10 trademarks to find other lines of symmetry.

3. Some trademarks can fit on themselves by turning the trademark. These trademarks have turn symmetry. Check to see if any of your 10 examples have turn symmetry. If not, find and cut one out of the yellow pages or the newspaper.

	Company's Name	Line of Symmetry Yes or No	More Lines of Symmetry Yes or No	Turn Symmetry Yes or No
1)				
2)				
3)				
4)				
5)				
6)				
7)				
8)				
9)				
10)				

Trademarks

Mathematics teaching objectives:

    . Find lines of symmetry.

    . Identify objects which have symmetry.

    . Recognize use of mathematics in art and advertising.

Problem solving skills pupils <u>might</u> use:

    . Classify objects.

    . Use diagrams.

Materials needed:

    . Telephone books or newpapers

Comments and suggestions:

    . The activity can be done with a whole group or small groups.
      Many businesses include their names as part of the trademark.
      The names seldom will have symmetry. Stress to the pupils that
      they are finding symmetry in just the shape (not the words) of
      the trademark.

    . An alternative to folding as a way of finding a line of symmetry
      is to use a mirror, shiny metal or piece of plastic. A shape has
      symmetry if it looks correct when a mirror is placed on a suspected
      line of symmetry.

    . Holding the trademark on a window or up to a bright light and then
      flipping it over allows pupils to identify those trademarks that
      have symmetry.

    . The more accurate terminology for turn symmetry is rotational
      symmetry. Most texts will use this latter term.

Answers:

    Answers will vary depending on trademarks chosen.

MISSION: SYMMETRY IN NATURE

Your mission--should you decide to take it--is to find 20 examples of symmetry in nature.

The examples may be:

a. actual samples from outside the school.
(Please do not destroy another person's property.)

b. pictures or illustrations from library books or textbooks.
(Don't tear them out.)

c. pictures or illustrations from magazines and newspapers.
(Cut them out if the magazine or newspaper is yours.)

d. written descriptions and/or demonstrations about examples that are still living.

Record your examples to make an attractive display.

Indicate a line of symmetry by a short, written description or by actually drawing a line of symmetry.

Be sure to finish your mission in a neat, orderly way or your teacher will self-destruct 10 seconds after this mission is completed.

PSM 81

Mission: Symmetry In Nature

Mathematics teaching objectives:

. Find lines of symmetry.

. Identify objects which have symmetry.

. Recognize that nature has mathematical aspects.

Problem-solving skills pupils might use:

. Collect data needed to solve a problem.

. Search printed material for needed information.

. Make and use physical models and diagrams.

. Search for other solutions.

Materials needed:

. Books, magazines, and the outdoors.

Comments and suggestions:

. This could be a project done over a period of several weeks.

Answers:

Answers will vary.

Grade 5

VI. DECIMALS

VI. DECIMALS

Decimals are usually a new topic in fifth grade. Because of this, problem solving with decimals needs to consider the teaching of concepts and operations. This section on decimals is quite developmental, using problem-solving when concepts are being learned and when background for operations is being given.

The development used is a decimals-only approach. No attempt is made to relate decimals and fractions. Several reasons for this approach seem valid:

1. The electronic calculator encourages use of decimals.

2. The movement to convert to the metric system suggests more emphasis on decimals than fractions.

3. Pupils may relate better to decimals as an extension of the whole-number system rather than conversions from and to fractions.

Using the Activities

These activities involve naming, ordering, comparing, equating, adding, and subtracting decimals. Several models are used: grids, number lines, and money*. No formal algorithms are given--only hands-on or discovery activities that lead toward the algorithms.

Because of the developmental approach, the activities are intended to be used in the order presented. Many of the activities can be finished quickly and will need supplementing from textbooks or other sources--as long as the material is consistent with the model used. The activities in the sections Estimation With Calculators and Story Problems could also be used with this section.

Extra materials needed for this section include:

. Orange and white Cuisenaire rods or comparable strips of paper

. Crayons or colored pencils

. Scissors

. Teacher-made deck of ten digit cards 0, 1, ..., 8, 9

---

*Money is a rather special model. Pupils often think of money in terms of whole numbers rather than decimals. Therefore, several concepts of decimals can be easily introduced by drawing on pupils' experiences with money.

TENTHS

This is one (1) square.

Use the marks to divide the square into 10 equal parts.

Each part is called one-tenth.

As a decimal this is written .1

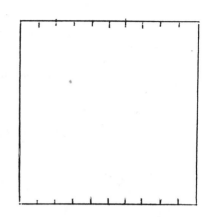

1. a. How many tenths are shaded?

   _____ -tenths

   b. How many tenths are not shaded?

   _____ -tenths

   c. Write both of these as a decimal.

   _____   _____

2. What part of each square is shaded? Write the answer in words and as a decimal.

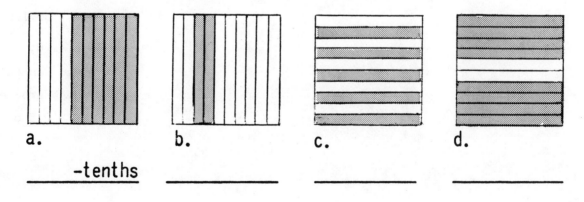

a.   b.   c.   d.

_____ -tenths   _____   _____   _____

_____   _____   _____   _____

Tenths

Mathematics teaching objectives:

. Indicate tenths using a grid model.

. Use word names and numeral names for decimals.

. Compare and order decimals in tenths.

Problem-solving skills pupils might use:

. Make and/or use a drawing.

. Be aware of other solutions.

Materials needed:

. None

Comments and suggestions.

. A grid is used as a model for writing tenths in both the word form
and numeral form.

. Pupils can work the two pages individually, or perhaps a transparency
of the first page could be used. Both sheets can be completed in
about 15 minutes.

. Pupils may need review on the meaning of $<$ and $>$.

. Emphasize problems g and h which name 0 as zero-tenths, .0 and 0.0,
and 1 as ten-tenths and 1.0. Some pupils may name ten-tenths as .10.

Answers:

1. a. Three-tenths
    b. Seven-tenths
    c. .3 and .7

2. Six-tenths, .6
    Two-tenths, .2
    Five-tenths, .5
    Eight-tenths, .8

3. a. .7
    b. Three-tenths
    c. Six-tenths
    d. .5
    e. .6, .7, .8, .9,
        or 1.0
    f. .0, .1, or .2
    g. Ten-tenths
    h. Zero-tenths,
        0.0 or .0

Shade each of these squares as indicated. Fill in the blanks.

a. Seven-tenths

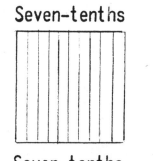

Seven-tenths = ._____

b. .3

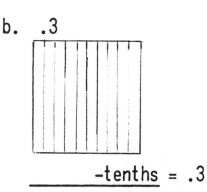

_____-tenths = .3

c. The decimal one-tenth
   greater than five-tenths.

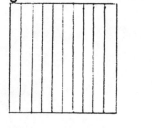

_____-tenths comes after
five-tenths

d. The decimal (in tenths)
   between .4 and .6

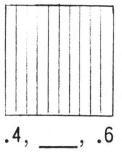

.4, ___, .6

e. Any decimal (in tenths)
   greater than .5

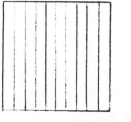

_____ > .5

f. Any decimal (in tenths)
   less than three-tenths

_____-tenths < three-tenths

g. The decimal (in tenths)
   greater than nine-tenths

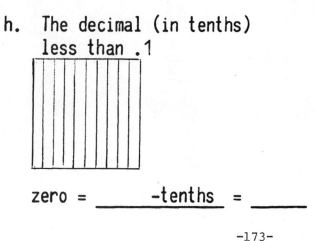

one = _____-tenths = 1.0

h. The decimal (in tenths)
   less than .1

zero = _____-tenths = _____

HUNDREDTHS

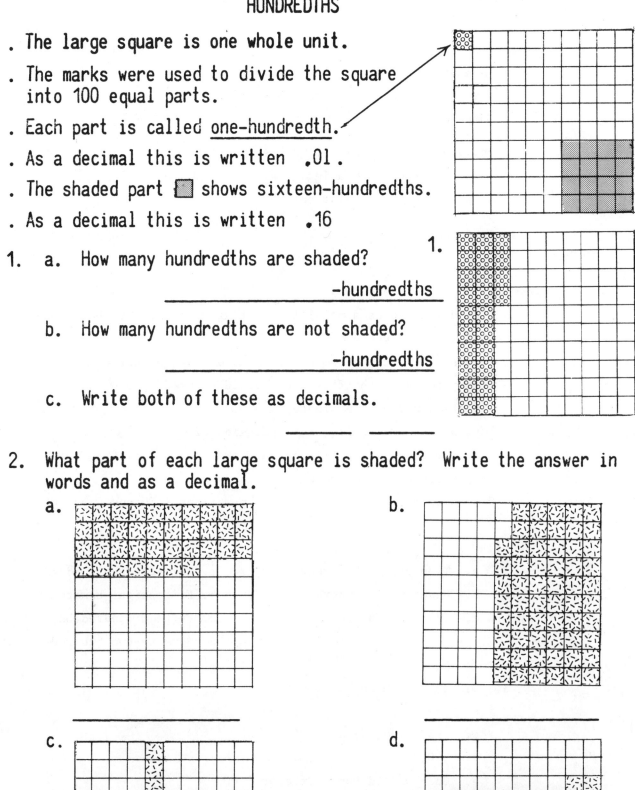

. The large square is one whole unit.

. The marks were used to divide the square into 100 equal parts.

. Each part is called <u>one-hundredth.</u>

. As a decimal this is written .01 .

. The shaded part ▒ shows sixteen-hundredths.

. As a decimal this is written .16

1. a. How many hundredths are shaded?

    _____ -hundredths

    b. How many hundredths are not shaded?

    _____ -hundredths

    c. Write both of these as decimals.

    _____   _____

2. What part of each large square is shaded? Write the answer in words and as a decimal.

    a.

    _____

    b.

    _____

    c.

    _____

    d.

    _____

Hundredths

Mathematics teaching objectives:

- Indicate hundredths using a grid model.
- Use word names and numeral names for decimals.
- Compare and order decimal in hundredths.

Problem-solving skills pupils <u>might</u> use:

- Make and/or use a drawing.
- Be aware of other solutions.

Materials needed:

- None

Comments and suggestions:

- A grid is used as a model for writing hundredths in both the word form and the numeral form.
- Pupils can work the two pages individually, or perhaps a transparency of the first page could be used. Both can be finished in about 15 minutes.
- Pupils may need review on the meaning of $<$ and $>$.
- Pupils may need help writing the word names, including the proper spelling and the use of the hyphen.
- Emphasize problems <u>e</u> and <u>f</u> which name <u>1 as 1.00</u> and <u>0 as .00</u> .

Answers:

1. a. Twenty-four hundredths
    b. Seventy-six hundredths
    c. .24 and .76

2. a. Thirty-seven hundredths, .37
    b. Fifty-eight hundredths, .58
    c. Fifteen hundredths, .15
    d. Seventy-two hundredths, .72

3. a. Thirty-three hundredths
    b. .51
    c. .93, .94, ..., 1.00
    d. .00, .10, ..., .09
    e. One-hundred hundredths or 1.00
    f. Zero hundredths or .00

3. Shade each of these squares as indicated. Fill in the blanks.

a. The decimal (one-hundredth) greater than 32-hundredths

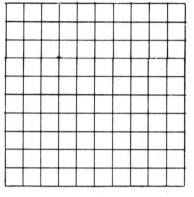

_____-hundredths comes after thirty-two hundredths

b. The decimal (in hundredths) between .50 and .52

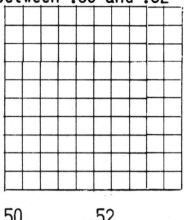

.50, _____ ,.52

c. Any decimal in hundredths greater than 92-hundredths

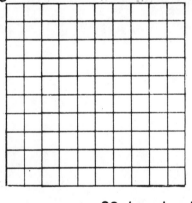

_____ > 92-hundredths

d. Any decimal (in hundredths) less than ten-hundredths

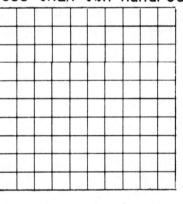

_____ < ten-hundredths

e. The decimal (in hundredths) greater than 99-hundredths

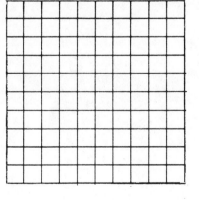

one = ____ -hundredths = 1.00

f. The decimal (in hundredths) less than .01

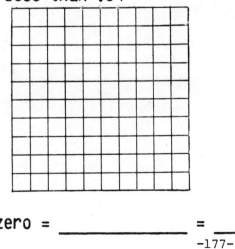

zero = _____ = _____

COVERING THE SQUARE

<u>GET</u> Orange and white rods

1. What is the <u>least</u> number
    of rods needed to com-
    pletely cover this square?

    _____

2. What is the <u>greatest</u>
    number of rods needed to
    completely cover this
    square?   _____

3. Which color rod represents
    .1 (one-tenth) of the
    square?

    _____

    .01 (one-hundredth)?

    _____

4. How much of the square can be covered by these rods? Write
    your answers as <u>decimals</u>. Some can be answered two ways.

    a. 10 white _____   e. 70 white _____
    b. 1 orange, 3 white _____   f. 100 white _____
    c. 43 white _____   g. 8 orange _____
    d. 6 orange, 8 white _____   h. 3 orange, 0 white _____

5. What is the [fewest] number of rods needed to cover the following
    parts of the square? Use only orange and white rods.

    a. .08 (eight-hundredths)   _____   e. .19   _____
    b. .30 (thirty-hundredths) _____   f. .70   _____
    c. .3 (three-tenths)   _____   g. .83   _____
    d. .03 (three-hundredths)   _____   h. .2   _____

<u>Covering</u> <u>The</u> <u>Square</u>

Mathematics teaching objectives:

. Use a grid model to indicate tenths and hundredths.

. Determine the simplest way to represent decimals.

Problem-solving skills pupils <u>might</u> use:

. Make and/or use a physical model.

. Be aware of other solutions.

Materials needed:

. 10 orange and about 25 white Cuisenaire rods or
1 cm x 10 cm and 1 cm x 1 cm strips of paper

Comments and suggestions:

. A grid is used as a model for tenths and hundredths. Pupils
use Cuisenaire rods or strips of paper to cover parts of the
grid.

. Several problems emphasize that 1., .2, etc. are equivalent
to .10, .20, etc.

. This activity could be done without the rods or strips, but
the "hands-on" placing of the rods on the grid is a valuable
part of the activity.

. After working problem 5, some pupils might realize that the
minimum number of rods can be determined by using the digit
in the tenths place for the number of orange rods and the
digit in the hundredths place for the number of white rods.
Ask them to explain this method to the class.

Answers:

1.	10	5.	a.	8 whites
2.	100		b.	3 oranges
3.	Orange, white		c.	3 oranges
4.	a. .1, .10		d.	3 whites
	b. .13		e.	10--1 orange and 9 whites
	c. .43		f.	7 oranges
	d. .68		g.	11--8 oranges and 3 whites
	e. .7, .70		h.	2 oranges
	f. 1.0, 1.00			
	g. .8, .80			
	h. .3, .30			

DIMES AND PENNIES

1. Joe is babysitting his sisters. His mother offers to pay him either $.80 per hour or .9 of a dollar per hour. Which offer should Joe accept? _____ Why? _____

   a. What coin is .01 of a dollar? _____

   b. What coin is .1 of a dollar? _____

   c. How does .10 of a dollar compare to .1 of a dollar? _____

2. Sally's father offers to pay her for the yard work. He says she can have the choice of 80¢ per hour or .8 of a dollar per hour.

   a. Which offer should Sally accept? _____

   b. Is Sally's father trying to be sneaky? _____

3. Sally and Joe were given this table to fill in. Only dimes and pennies are allowed. They had to decide whether Part A was
   $<$, $>$, or $=$ to Part B.

	A			B	
Coins	Decimal name		Decimal name	Coins	
2 dimes	.2	=	.20	20 pennies
2 dimes 3 pennies		$<$	.3	
	.5	___		4 dimes 6 pennies
	.62	___	.6	
	.80	___	.8	
	.91	___	.9	
	.4	___	.40	
	.68	___	.7	

Dimes And Pennies

Mathematics teaching objectives:

- Use a money model to indicate tenths and hundredths.
- Show that .1, .2, etc. are equivalent to .10, 20, etc.
- Compare decimals.

Problem-solving skills pupils _might_ use:

- Make and use a table.
- Use a model.
- Solve a similar but related problem.

Materials needed:
- None

Comments and suggestions:

- Money is a practical application and model for decimals. Special care is needed to interpret 39¢ as .39 of a dollar, $1.98 as 1.98 and 1 dime as .10 .

- This activity emphasizes the problem-solving skill _use a model_. To compare .8 to .24, pupils can compare .8 of a dollar to .24 of a dollar. 8 dimes is more than 2 dimes and 4 pennies, so .8 > .24. If pupils are asked _why_ the model helped to solve the problem, they might answer, "Because we know more about dimes and pennies." Using a familiar model can help to solve unfamiliar problems.

- The primary objective of this activity is for pupils to understand that .1, .2, etc. can be written as .10, .20, etc. and vice versa. The addition or deletion of zeros on the "end" of decimals does not affect the value, only the way the decimal is named. Remind pupils that a deletion of zeros for whole numbers _does_ affect the value.

Answers:

1. Joe: .9 is the same as 90¢   2. Sally: a. Either--they are the same.
   a. Penny   b. Dime   b. Answers will vary.
   c. Same

Coins	Decimal Name		Coins	Decimal Name
	.23	<		3 dimes
5 dimes		>	.46	
6 dimes 2 pennies		>		6 dimes
80 pennies		=		8 dimes
9 dimes 1 penny		>		9 dimes
4 dimes		=		40 pennies
6 dimes 8 pennies		<		7 dimes

(Some might answer 30 pennies.)

DECIMAL DIGITS
(Ideas For Teachers)

Have pupils draw the frames shown below on a piece of paper.
Show pupils ten digit cards labeled 0-9. Explain that you will
draw five digits, one at a time. Pupils are to place the digits
in the frames trying to make the largest possible decimal. Once
placed, a digit may not be changed. One of the digits may be
rejected and written in the circle.

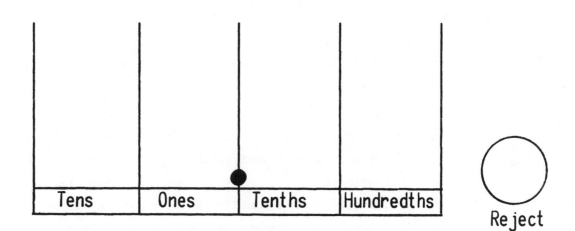

The following are variations for the activity.

a. Use the same five digits to make the smallest possible
   decimal.

b. Use the same five digits to make the decimal closest
   to 55.55 .

c. Repeat the activity but replace each digit before the
   next one is drawn.

d. Choose or draw any four digits. Make as many different
   decimals as possible. Arrange the decimals in order
   from largest to smallest.

Decimal Digits

Mathematics teaching objectives:
. Use place value concepts to form a specified number.
. Use probability concepts in making "best" choices.

Problem-solving skills pupils might use:
. Make decisions based upon data.
. Be aware of other solutions.

Materials needed:
. Teacher-made deck of ten digit cards 0, 1, 2, ..., 8, 9

Comments and suggestions:
. This activity works well as a short warm-up at the beginning of class or as a wind-up for the last few minutes of the class.
. A particular goal, such as largest number, is set. Digits are drawn, one at a time, while pupils write the digits in their place-value frames or in the reject circle. No changes are allowed.
. After several games, pupils can share the strategies they have developed. Questions can be used to bring these strategies out. "If the first draw is a 9 where would you put it? Why? What if it were a 0? 8? 3? 5? "
. The chance factor introduced by drawing digits provides all pupils with an opportunity to get the best answer.
. For other variations, read "Digit Draw Activities" located in the Whole Number Drill and Practice section.

Answers:
Answers will vary according to the digits drawn and placement by the pupils. For example, if the digits were 3, 6, 7, 2, and 9, two different pupils might make these numbers:
(1) 76.32   or   (2) 96.23 .

PATTERNS IN A HUNDREDTHS TABLE

Rene found a system to label these squares. Continue the
pattern to label the other squares.

.01	.02	.03					.08		.1
				.15	.16			.19	
								.29	
	.32		.34			.37			

Pretend that Rene's grid were extended. Fill in the missing
decimals.   Look for patterns.

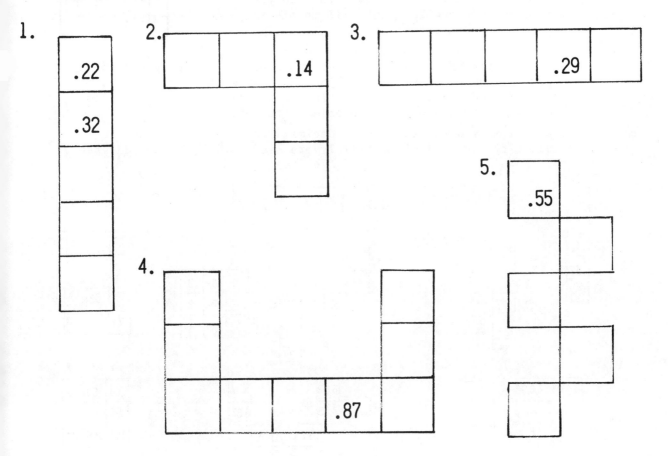

1.

.22
.32

2.

		.14

3.

			.29	

4.

5.

.55

.87

) PSM 81

Patterns In A Hundredths Table

Mathematics teaching objectives:

. Use patterns to order and compare decimals.

Problem-solving skills pupils _might_ use:

. Make and/or use a drawing or diagram.
. Look for and use patterns.
. Break a problem into parts.

Materials needed:

. None

Comments and suggestions:

. Have pupils complete the grid at the top. Be sure pupils don't confuse this grid with the hundredths grid used on previous pages.

. Some pupils will label .1, .2, .3, and .4 as .10, .20, .30, and .40. Review this concept from previous lessons.

. Questions can be used to help students focus on the patterns. "What pattern do you see in the first row? Does that pattern work in the other rows? What pattern is in the first column? Is it the same in the other columns? Is there a diagonal pattern? Is it the same for diagonals in other directions?" (↗ versus ↘)

. Problems 2, 3, and 7 can be copied directly from the grids. See the drawings to the right. Some pupils might fill in missing parts in the grids.

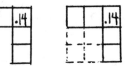

. Pupils can use the basic patterns to solve problems 1, 2, 3 and 4. If they get stuck on 5 or later problems, ask if they can solve part of the problem. In 5 maybe they can work down from .55 to get .75 and .95 before attempting to fill in the other squares. Emphasize that breaking a problem into parts usually helps.

. Unlike many situations where decimals represent measures of quantity, these decimals simply are used as labels (or addresses).

Answers:

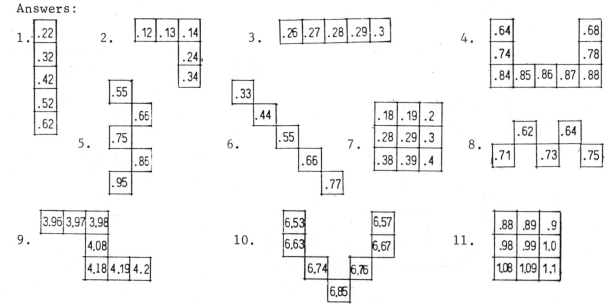

-186-

Patterns in a Hundredths Table (cont.)

For some of these, pretend the grid goes on.

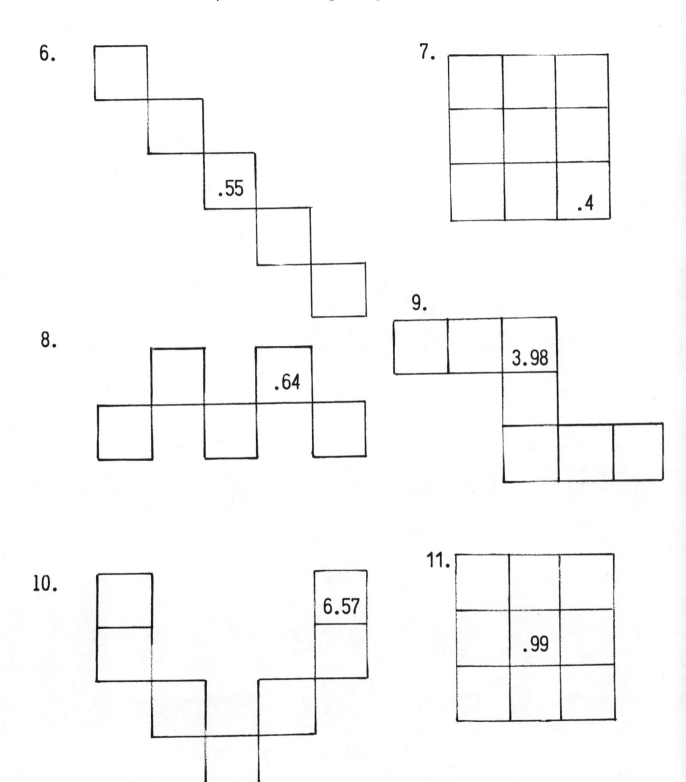

6.

   .55

7.

   .4

8.

   .64

9.

   3.98

10.

   6.57

11.

   .99

AMAZING DECIMALS

Find a path through the decimal maze.
Always move to a larger decimal.

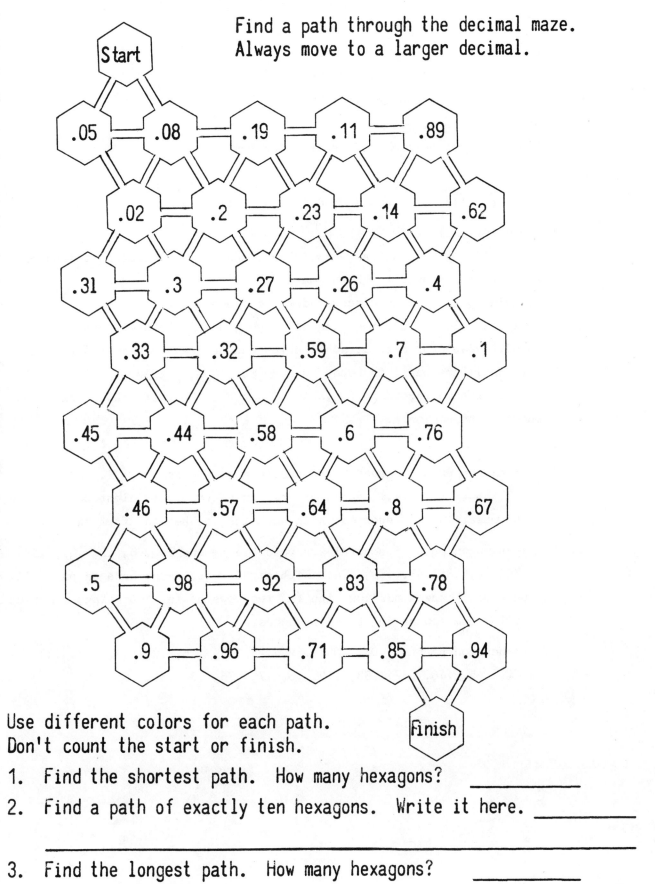

Use different colors for each path.
Don't count the start or finish.

1. Find the shortest path. How many hexagons? _____

2. Find a path of exactly ten hexagons. Write it here. _____

_____

3. Find the longest path. How many hexagons? _____

<u>Amazing</u> <u>Decimals</u>

Mathematics teaching objectives:

. Compare decimals.

. Recognize that .1, .2, etc., are equivalent to .10, .20, etc.

Problem-solving skills pupils <u>might</u> use:

. Guess and check.

. Search for or be aware of other solutions.

Materials needed:

. Pens, pencils, or crayons of different colors

Comments and suggestions.

. Pupils likely will use a guess, check, and refine approach to finding appropriate paths through the maze.

. Certain spots will be discovered to be deadends, e.g., .96.

. Pupils should not use the pens or crayons until the best path has been found. Pieces of corn or small chips can be used to keep track of a path until the pupil is ready to draw it. Some pupils might prefer to keep a written record so they can adjust their paths without going back to the beginning.

Answers:

1. 8-hexagon path -- several are possible -- One is .08, .2, .27, .32, .58, .64, .83, .85. Pupils may recognize that only eight rows are in the maze. So the shortest path contains one hexagon from each row.

2. 10-hexagon path -- several are possible -- One is .08, .2, .27, .59, .7, .76, .8, .83, .85, .94. With eight rows, a ten-hexagon path will have one hexagon per row with (a) two rows having two hexagons marked or (b) one row having three hexagons.

3. 23-hexagon path -- .05, .08, .19, .2, .23, .26, .27, .3, .31, .33, .44, .45, .46, .57, .58, .59, .6, .7, .76, .8, .83, .85, .94

THE WHOLE THING

Solve these problems.   Try to find an easy way.

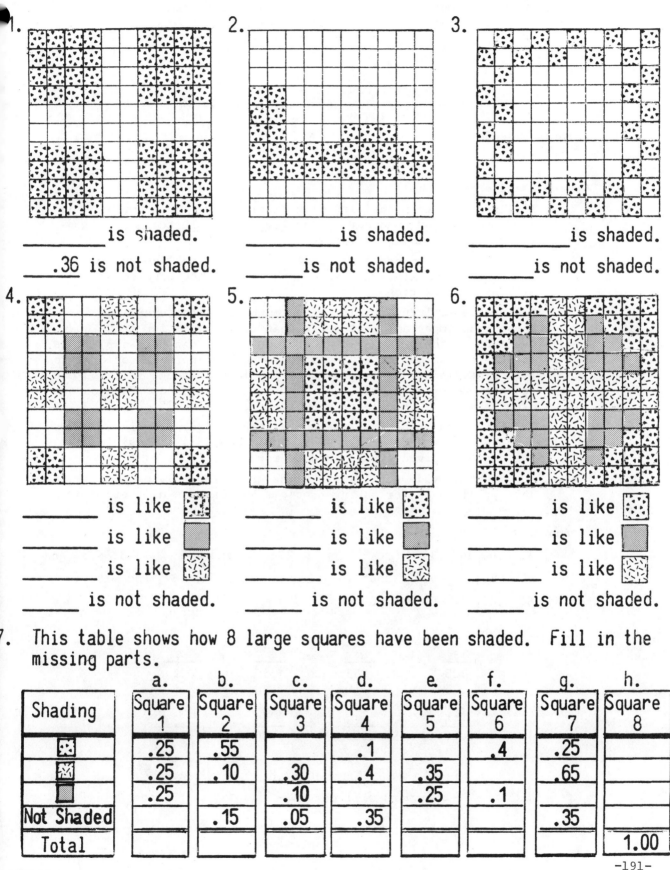

1. _____ is shaded.

   __.36__ is not shaded.

2. _____ is shaded.

   _____ is not shaded.

3. _____ is shaded.

   _____ is not shaded.

4. _____ is like ▨

   _____ is like ▨

   _____ is like ▨

   _____ is not shaded.

5. _____ is like ▨

   _____ is like ▨

   _____ is like ▨

   _____ is not shaded.

6. _____ is like ▨

   _____ is like ▨

   _____ is like ▨

   _____ is not shaded.

7. This table shows how 8 large squares have been shaded. Fill in the missing parts.

Shading	a. Square 1	b. Square 2	c. Square 3	d. Square 4	e. Square 5	f. Square 6	g. Square 7	h. Square 8
▨	.25	.55		.1		.4	.25	
▨	.25	.10	.30	.4	.35		.65	
▨	.25		.10		.25	.1		
Not Shaded		.15	.05	.35			.35	
Total								1.00

-191-

© PSM 81

<u>The</u> <u>Whole</u> <u>Thing</u>

Mathematics teaching objectives:

. Use a grid model to indicate tenths and hundredths.
. Recognize that 1.00 represents a whole unit.
. Develop background for adding and subtracting decimals.

Problem-solving skills pupils <u>might</u> use:

. Look for and use patterns.
. Search for or be aware of other solutions.

Materials needed:

.None

Comments and suggestions:

. Pupils can do this activity with little help. Some emphasis on
  <u>1.00 meaning 100 hundredths</u>, and <u>1.0 meaning 10 tenths</u>, and
  <u>1.00 = 1.0 = 1</u> may be helpful.

. For the first six problems, some pupils will realize that what
  is shaded and what is not shaded must equal 100-hundredths and
  will use subtraction.

. The problem-solving skills can be emphasized in several different
  ways. The important pattern is that the answers for each square
  must add up to 100-hundredths or 1.00. They can then apply this
  pattern to the bottom of the page. Another approach would be to
  give hints individually <u>only</u> when pupils get stuck. "If you can't
  get the bottom problems, look back at the top ones. Can you see any
  clues? What can you learn from the top problems?"

. The first four tables at the bottom have unique answers. Tables
  5, 6, and 8 have several answers and table 7 is impossible because
  the decimals given already add to more than 1.00.

. You might provide pupils with blank grids for solving some of the
  last problems or for making problems and designs of their own.

Answers:

1. .64 and .36   2. .29 and .71   3. .32 and .68

4. .16, .16, .20, and .48   5. .16, .36, .32, and .16

6. .40, .24, .36, and .00

7. Tables:

a.	1	b.	2	c.	3	d.	4
	.25		.55		.55		.1
	.25		.10		.30		.4
	.25		.20		.10		.15
	.25		.15		.05		.35
	1.00		1.00		1.00		1.00

e.	5	f.	6	h.	8	g.	7
	Answers will vary						Impossible

MAKING CENTS OUT OF MONEY

Money--dollars and cents--is a good application of decimals. You should be accurate with calculations but it is also important to be a good estimator.

Ben is helping his dad do the shopping. Estimate to see if they can buy these five items. They have only $10.00.

1 dozen eggs	$ .89	cheese	$1.99	ham	$3.95
lunch meat	$1.07	milk	$1.59		

Do you think they can buy these five items? _____

Ben thought this way:
$.89 is almost $1.00, so

$ .89 ⟶ $ 1.00
$1.07 ⟶ $ 1.00
$3.95 ⟶ $ 4.00
$1.99 ⟶ $ 2.00
$1.59 ⟶ $ 2.00
_____
$10.00 should be enough

Now you be the best shopper. Write your estimates in the blanks. Then circle the best estimate.

1. $1.89 $3.93 $2.89 $3.79

___ ___ ___ ___

$9.00 $11.00 $13.00

2. $3.09 $1.89 $.49 $2.61

___ ___ ___ ___

$6.00 $7.00 $8.00

3. $2.11 $3.14 $1.09 $4.19

___ ___ ___ ___

$10.00 $11.00 $14.00

4. $.83 $1.19 $4.09 $2.87 2 of these

___ ___ ___ ___

$8.00 $10.00 $12.00

5. 2 of these $5.89 $2.98 3 of these

___ ___ ___

$16.00 $19.00 $21.00

6. $5.79 $3.09 4 of these

___ ___

$8.00 $18.00 $22.00

PSM 81

Making Cents Out of Money

Mathematics teaching objectives:

. Use decimals in solving money problems.

. Estimate money amounts using rounding procedures.

Problem-solving skills pupils might use:

. Make reasonable estimates.

Materials needed:

. None

Comments and suggestions;

. Many pupils would rather compute the answer exactly than estimate. The fear of being wrong inhibits them. A positive class atmosphere and a series of activities in estimation might help. The section on Estimation With Calculators can be used.

. This activity gives pupils practice in either rounding up or rounding down to a convenient amount and then using those amounts to get an estimate. Other activities are important and needed to emphasize the importance of estimation, especially while shopping.

. After all estimates have been made, pupils may want to find the exact answer to see how close it is to the estimate. They can do this by hand or with a calculator. Insisting that they check every estimate with computation might reduce pupil motivation to estimate.

Answers:

1. $13.00

2. $8.00

3. $10.00 Although $10.00 is the estimate most pupils will get for this problem, $11.00 is closer to the actual answer. Discuss how rounding down for each price tag caused the estimate to be inaccurate.

4. $12.00

5. $21.00

6. $18.00

TENTHS ON THE NUMBER LINE

0 units ——|

Needed: 1 orange rod and 12 white rods

1. a. How many white rods make the same length
        as an orange rod? _____

    b. One white rod is <u>one-tenth</u> of an orange rod.

    c. This is written as <u>.1</u> .

2. Lay the orange rod along the line at the right.

    a. Start with one end at the zero mark.
    b. Mark the other end. Label the end <u>1 unit</u>.
    c. Place the orange rod at 1 .
    d. Mark the other end. Label the end <u>2 units</u>.

3. Use a white rod.

    a. Start at zero.
    b. Mark the other end. Label the end as <u>.1</u> .
    c. Place the white rod at <u>.1</u> .
    d. Mark the other end. Label the end <u>.2</u> .
    e. Continue until you have marked and labeled .9 .

    f. Mark another length of a white rod.

    g. At what already marked point are you? _____
    h. How many tenths have you marked? _____
    i. This point has several names:
            one, ten-tenths, 1, or 1.0 .

4. Continue marking tenths from 1 to 2 .

    a. How many tenths are between 0.0 and 1.0? ____
    b. How many tenths are between 1.0 and 2.0? ____
    c. How many tenths are between 0.0 and 2.0? ____

5. Mark tenths as far past 2.0 as you can.

6. If the line were longer,
    a. how many tenths would be between 0.0 and 4.0? ____
    b. how many tenths would be between 3.0 and 5.0? ____
    c. how many tenths would be between 6.5 and 7.5? ____
    d. how many tenths would be between 8.3 and 9.6? ____

PSM 81

Tenths On The Number Line

Mathematics teaching objectives:

- Use a number line as a model for tenths.
- Locate and name points (in tenths) on a number line.
- Name 0, 1 and 2 as 0.0, 1.0 and 2.0 respectively.
- Develop background for subtracting decimals.

Problem-solving skills pupils might use:

- Use a physical model.
- Look for and use a pattern.

Materials needed:

- Orange and white Cuisenaire rods or 10 cm and 1 cm strips.

Comments and suggestions:

- To ensure that all pupils mark their lines properly, this activity is best done as a teacher directed activity with a demonstration being done on the overhead projector while pupils work at their desks.
- After completion of the labeling, have pupils point to each point on the number line as they orally read the name for each point.
- Mention again the several names for the points 0, 1 and 2.
- After working problem 6, have pupils explain how they decided on the answers. Did they use the white rods to mark on an extended imaginary line? Did they use a pattern? Did they compute? Count? Sharing strategies can help pupils see that there are many correct ways of solving a problem.

Answers:

1. a. 10

3. g. 1   h. 10

4. a. 10   b. 10   c. 20

6. a. 40   b. 20   c. 10   d. 13

NUMBER LINE DECIMALS

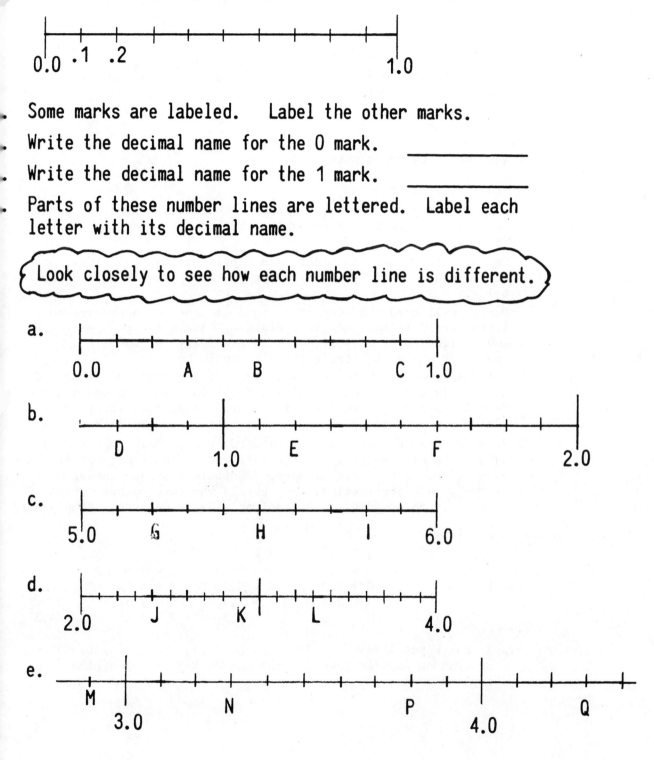

mber lines can help us understand decimals. This segment is
vided into 10 equal parts.

0.0 .1 .2   1.0

. Some marks are labeled.   Label the other marks.

. Write the decimal name for the 0 mark. _____

. Write the decimal name for the 1 mark. _____

. Parts of these number lines are lettered. Label each
   letter with its decimal name.

Look closely to see how each number line is different.

a.
0.0   A   B   C 1.0

b.
   D   1.0   E   F   2.0

c.
5.0   G   H   I   6.0

d.
2.0   J   K   L   4.0

e.
M   3.0   N   P   4.0   Q

Number Line Decimals

Mathematics teaching objectives:
- Use a number line as a model for tenths.
- Name and locate points (in tenths) on a number line.
- Name 0 and 1 as 0.0 and 1.0 respectively.
- Use multiples of decimals.

Problem-solving skills pupils might use:
- Make and/or use a drawing.
- Make reasonable estimates.
- Look for and use patterns.

Materials needed:
- Centimetre ruler or note card

Comments and suggestions:
- This activity uses number lines as a model for naming and locating tenths. Be sure pupils recognize that the unit length changes from problem to problem.
- On the second page you might have pupils estimate the location of the wanted decimal before using the more exact method described below.
- Pupils will need the edge of a piece of paper or note card to locate the points. The given distance can be marked and duplicated as many times as is necessary to find the point. A centimetre ruler also could be used.
- Patterns occur both in the multiples (seeing that .0, .4 and .8 are evenly spaced) and in solving problems like M in 4e (see that P in 4e is 3.8 by seeing it is .8 ahead of 3, then study the relationship of P to 4.0 and use that pattern to decide M is 2.9).
- Problem i can be used to start a discussion. Most pupils will assume the zero point is at the left end of the line, but it could be anywhere to the left of zero. There is a unique answer only if two points are labeled. Any pupils who labeled some point .0 or .1 ... or decided i was unsolvable as is deserve a compliment!

Answers:

2.	0.0	4. a.	A = .3,	B = .5,	C = .9
		b.	D = .7,	E = 1.2,	F = 1.6
3.	1.0	c.	G = 5.2,	H = 5.5,	I = 5.8
		d.	J = 2.4,	K = 2.9,	L = 3.3
		e.	M = 2.9,	N = 3.3,	P = 3.8 Q = 4.3

5. To check pupil answers, draw the answers on a transparency or tracing paper. Then lay the answer key on top of the pupil's page.

5. For these number lines, find and mark the named decimals.

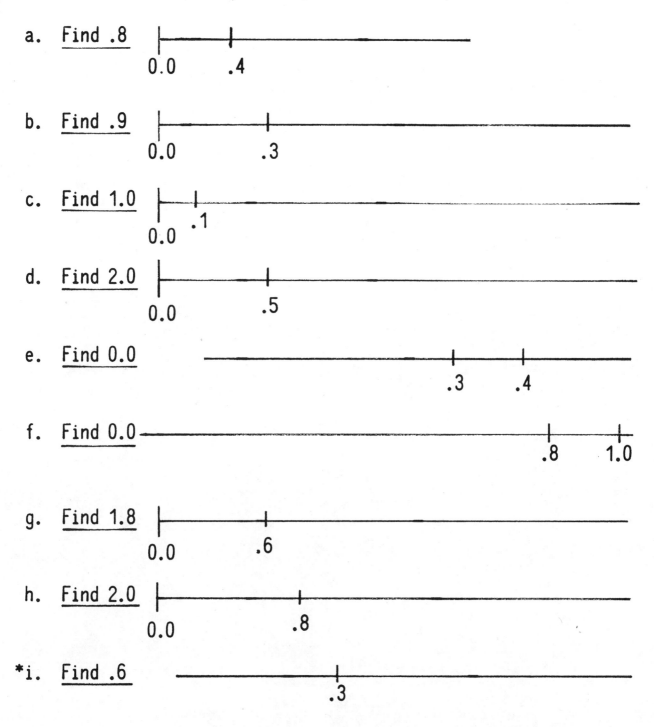

a. Find .8

   0.0   .4

b. Find .9

   0.0   .3

c. Find 1.0

   0.0 .1

d. Find 2.0

   0.0   .5

e. Find 0.0

             .3   .4

f. Find 0.0

               .8   1.0

g. Find 1.8

   0.0   .6

h. Find 2.0

   0.0   .8

*i. Find .6

         .3

ADDING ON THE NUMBER LINE

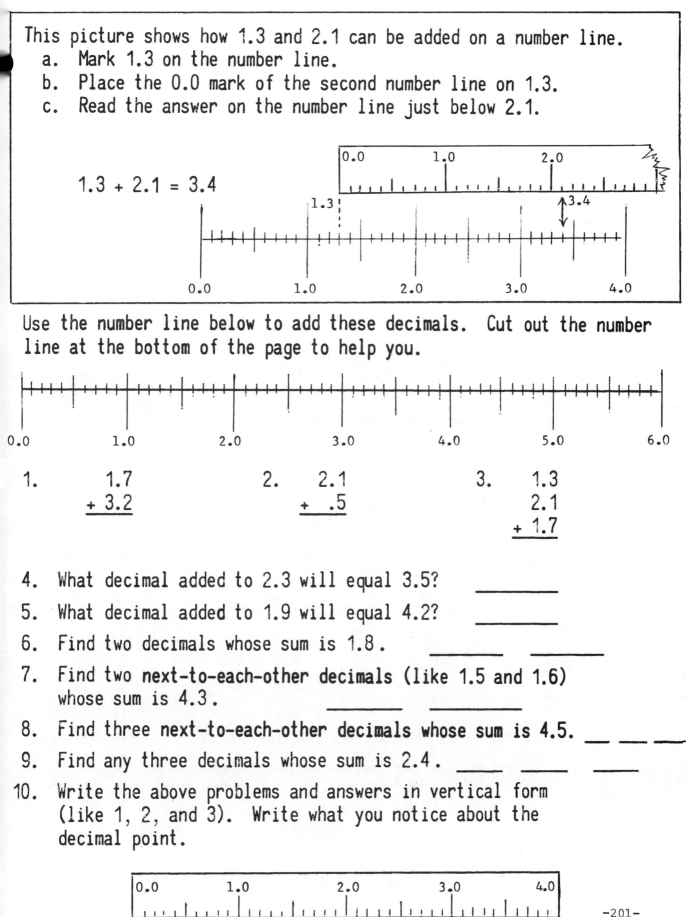

This picture shows how 1.3 and 2.1 can be added on a number line.
  a. Mark 1.3 on the number line.
  b. Place the 0.0 mark of the second number line on 1.3.
  c. Read the answer on the number line just below 2.1.

1.3 + 2.1 = 3.4

Use the number line below to add these decimals. Cut out the number line at the bottom of the page to help you.

1.	1.7	2.	2.1	3.	1.3
	+ 3.2		+ .5		2.1
					+ 1.7

4. What decimal added to 2.3 will equal 3.5?   _____

5. What decimal added to 1.9 will equal 4.2?   _____

6. Find two decimals whose sum is 1.8.   _____ _____

7. Find two **next-to-each-other decimals** (like 1.5 and 1.6) whose sum is 4.3.   _____ _____

8. Find three **next-to-each-other decimals whose sum is 4.5.** __ __ __

9. Find any three decimals whose sum is 2.4. ____ ____ ____

10. Write the above problems and answers in vertical form (like 1, 2, and 3). Write what you notice about the decimal point.

Adding On The Number Line

Mathematics teaching objectives:

- Add decimals (tenths) on a number line.
- Discover a system for adding decimals (tenths) without using a number line.

Problem-solving skills pupils *might* use:

- Use a physical model.
- Work backwards.
- Generalize from several examples.

Materials needed:

- Scissors to cut out the number-line rulers
- Demonstration rulers for use on the overhead projector

Comments and suggestions:

- Demonstrate addition of decimals on a number line using the overhead or large paper models.
- Have pupils cut out their own rulers and use them to solve the first three problems. Pupils might work backwards to solve problems 4-9 by starting with the sum and measuring the distance between the given addend and the sum. Some pupils will solve problems 4-9 without having to use the rulers.
- A result of this activity will be a feeling for the "rule" of adding decimals, that is, "keep the decimal points lined up."
- Explain to pupils that consecutive (or next-to-each-other) decimals do not exist. The term is used here to indicate decimals like 3.5, 3.6, 3.7, etc.

Answers:

1. 4.9   2. 2.6   3. 5.1   4. 1.2   5. 2.3

6. 1.2 + .6 and .3 + 1.5 are two possibilities.

7. 2.1 and 2.2

8. 1.4, 1.5, and 1.6

9. .7 + .8 + .9 and 1.4 + .8 + .2 are two possibilities.

10. We could have just added these if we had kept the decimal points lined up.

DECIMAL PERIMETERS

Cut out the tenths ruler at the bottom of the page. Use it to measure and record the lengths of each side. Find the perimeter of each figure.

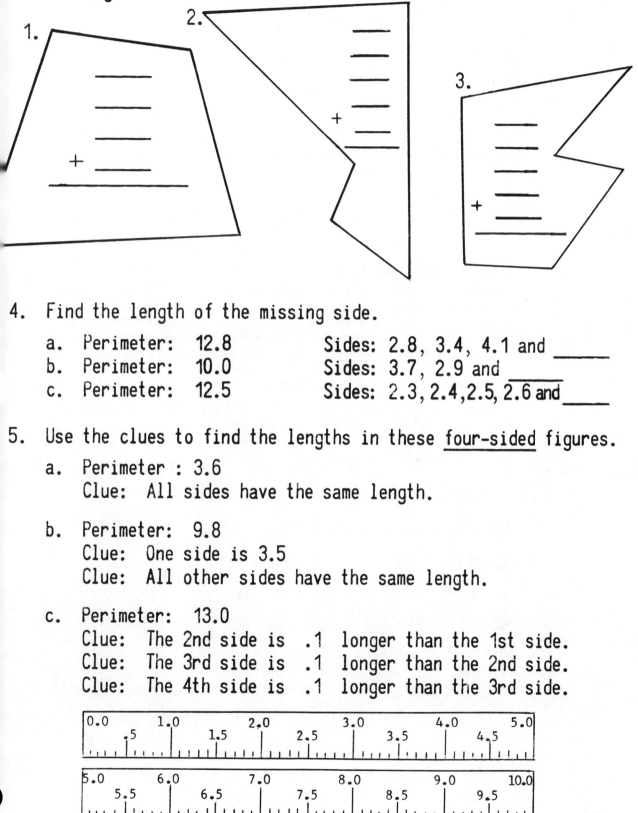

4. Find the length of the missing side.

    a. Perimeter: 12.8   Sides: 2.8, 3.4, 4.1 and _____
    b. Perimeter: 10.0   Sides: 3.7, 2.9 and _____
    c. Perimeter: 12.5   Sides: 2.3, 2.4, 2.5, 2.6 and _____

5. Use the clues to find the lengths in these <u>four-sided</u> figures.

    a. Perimeter : 3.6
        Clue: All sides have the same length.

    b. Perimeter: 9.8
        Clue: One side is 3.5
        Clue: All other sides have the same length.

    c. Perimeter: 13.0
        Clue: The 2nd side is .1 longer than the 1st side.
        Clue: The 3rd side is .1 longer than the 2nd side.
        Clue: The 4th side is .1 longer than the 3rd side.

Decimal Perimeters

Mathematics teaching objectives:

   . Add decimals (tenths).
   . Find perimeters of polygons.

Problem-solving skills pupils **might** use:

   . Use a physical model.
   . Guess and check.
   . Make a drawing or diagram.
   . Work backwards.

Materials needed:

   . Scissors to cut out the number line ruler
   . Tape to attach the parts

Comments and suggestions:

   . Pupils may think of each polygon as a single line that has been
     bent. On problems 1, 2 and 3 each side must be measured separately.
     To find the perimeter (sum) pupils can rotate their ruler around
     the shape until they have the total length.

   . If pupils are stuck on problem 4, ask if it would help to make
     a drawing. Some might start a polygon with the given sides and
     discover they cannot complete the polygon and stay within the
     12.8 units. Others might work backwards and draw a line 12.8
     units long, measure off the given sides one at a time and measure
     what is left. Pupils who tried strategies that didn't work can
     also contribute to the discussion and understanding of the problem.

   . The problems in number 5 can be solved by guessing and checking.
     No multiplication or division is needed.

Answers:

   1. $2.5 + 1.6 + 2.1 + 2.9 = 9.1$

   2. $1.1 + .6 + 2.4 + 2.3 + 3.0 = 9.4$

   3. $1.8 + 1.9 + 1.3 + .8 + 1.4 + 1.0 = 8.2$

   4. a. 2.5   b. 3.4   c. 2.7

   5. a. .9,   .9,   .9,   .9
       b. 3.5,   2.1,   2.1,   2.1
       c. 3.1,   3.2,   3.3,   3.4

SUBTRACTING ON THE NUMBER LINE

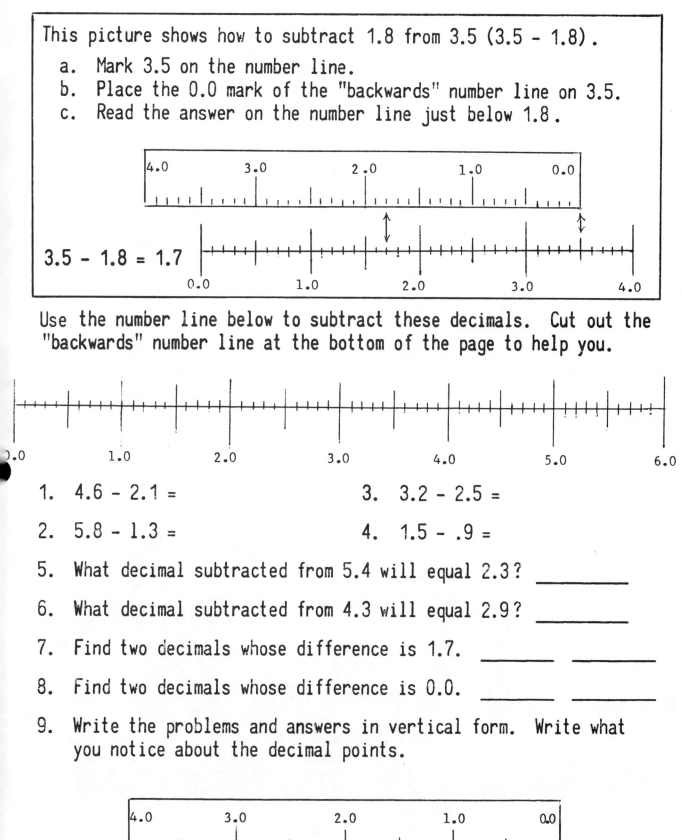

This picture shows how to subtract 1.8 from 3.5 (3.5 - 1.8).

a. Mark 3.5 on the number line.
b. Place the 0.0 mark of the "backwards" number line on 3.5.
c. Read the answer on the number line just below 1.8.

3.5 - 1.8 = 1.7

Use the number line below to subtract these decimals. Cut out the "backwards" number line at the bottom of the page to help you.

1. 4.6 – 2.1 =

2. 5.8 – 1.3 =

3. 3.2 – 2.5 =

4. 1.5 – .9 =

5. What decimal subtracted from 5.4 will equal 2.3? _____

6. What decimal subtracted from 4.3 will equal 2.9? _____

7. Find two decimals whose difference is 1.7. _____ _____

8. Find two decimals whose difference is 0.0. _____ _____

9. Write the problems and answers in vertical form. Write what you notice about the decimal points.

PSM 81

Subtracting On The Number Line

Mathematics teaching objectives:

- Subtract decimals (tenths) on a number line.
- Discover a system for subtracting decimals (tenths) without using a number line.

Problem-solving skills pupils _might_ use:

- Use a physical model.
- Work backwards.
- Study the solution process.

Materials needed:

- Scissors to cut out the "backwards" ruler
- Demonstration ruler for use on the overhead projector.

Comments and suggestions:

- Use the overhead projector to demonstrate this interpretation of subtraction on a number line. (There are other important interpretations!)
- A result of this activity will be a feeling for the "rule" of subtracting decimals, that is, "keep the decimal points lined up."

Answers:

1. 2.5
2. 4.5
3. .7
4. .6
5. 3.1
6. 1.4
7. Answers will vary. 2.9-1.2 is one possibility.
8. Answers will vary. Any decimal subtracted from itself will work.
9. We could have just subtracted these if we had kept the decimal points lined up.

THE DECIMAL WEIRD RULER

(Ideas for Teachers)

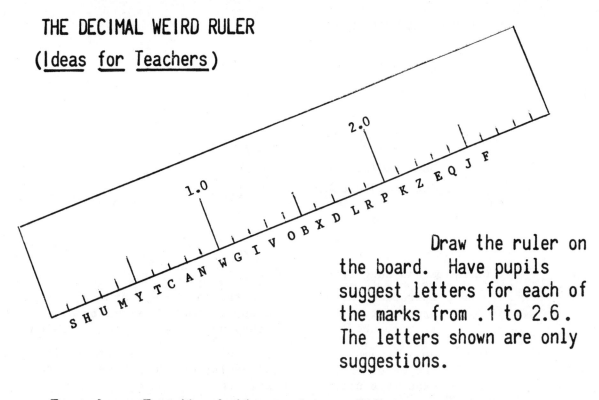

Draw the ruler on the board. Have pupils suggest letters for each of the marks from .1 to 2.6. The letters shown are only suggestions.

Example: For the letters shown, PAT has a length of 2.0 + .8 + .6 = 3.4.

Questions which could be asked:

1. Whose name has the greatest length when you add the lengths of each letter?

2. Which is longer (according to this ruler): yard or metre?

3. Can you find a word whose length is exactly 5 units?

4. Using this ruler what 3-letter word has the greatest length?

5. Have pupils create their own questions.

CHALLENGE

Find the item in the classroom with the greatest length.

The Decimal Weird Ruler

Mathematics teaching objectives:
  . Read a number line.
  . Add decimals (tenths).

Problem-solving skills pupils might use:
  . Make and/or use a drawing.
  . Guess and check.

Materials needed:
  . None

Comments and suggestions:
  . Introduce the activity on the chalkboard or with the overhead
    projector. You can use the letters as shown or allow pupils to
    arrange the letters as they want. (Beware of"four-letter" words.)
  . Do an example together, then allow pupils time to investigate
    the other questions. An additional search is to find a word worth
    exactly 6 units, 7 units, etc.
  . Writing a word vertically helps make the addition easier.
  . Conclude with a class discussion to allow pupils to share their
    results.

Answers:

  1. Answers will vary.

  2. M   .4   Y   .5
      E   2.3   A   .8   METRE is 2.6 units longer.
      T   .6   R   1.9
      R   1.9   D   1.7
      E   2.3   4.9
          7.5

  3. Answers will vary. ROD is one such word.

  4. FED is 6.6 units long. (Pupils may find a longer word.)

DECIMAL PATHS

Draw a path so the numbers along the path add to the number in the circle. The path can only go through the open gates.

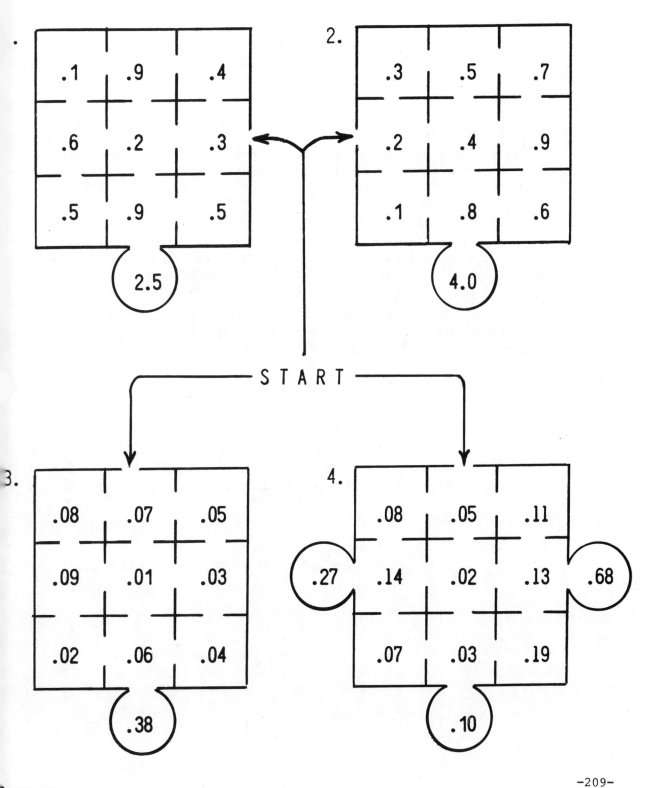

© PSM 81

Decimal Paths

Mathematics teaching objectives:

. Add decimals (tenths and hundredths).

Problem-solving skills pupils might use:

. Guess and check.

. Break a problem into parts.

Materials needed:

. None

Comments and suggestions:

. Many pupils will guess, check, and refine for these problems.
Try not to give hints unless pupils are really stuck. Hints
could include, "Was the path you tried too big? Could you cut
out some numbers? Can you find any combination of numbers that
adds up to what you need? ..."

. Some may realize that for each problem the beginning and end
numbers are determined so the problem is to solve the middle part.
For example, in problem 1, .3 and .9 must be used, so the problem
is to find a path totaling 2.5 - 1.2 (1.3 units).

. The .68 path in the last problem can be done only by passing
through .05 twice. Pupils may ask if this is O.K. Note that
the directions do not prohibit such moves.

Answers:

1. $2.5 = .3 + .2 + .6 + .5 + .9$

2. $4.0 = .2 + .3 + .5 + .7 + .9 + .6 + .8$

3. $.38 = .07 + .08 + .09 + .01 + .03 + .04 + .06$

4. $.27 = .05 + .08 + .14$

   $.10 = .05 + .02 + .03$
   $.68 = .05 + .08 + .14 + .07 + .03 + .02 + .05 + .11 + .13$

Grade 5

VII. PROBABILITY

Experimental probability provides
an excellent opportunity to emphasize
these problem solving skills:
collect data, make a table,
use a model, make predictions
and in general, guess, check
and refine. As students suggest
ways of solving the problems
(or as they describe how they
solved the problems) their
answers can be translated into
standard problem solving phrases.
For example, "Keep track of what
you get" can be verbalized by the teacher as "Make a table." Specific
suggestions for emphasizing the problem solving skills are given in the
comments and suggestions for each page.

Working with experimental probability in grades 5 and 6 is also instruc-
tionally sound. Usually students are not ready for any formal probability
like "The probability of rolling a sum of 7 is $\frac{7}{36}$," until somewhere between
the seventh and tenth grades. Experimental probability provides some background for later work with formal probability and, more importantly, it
provides experience with a method of solving problems that is very useful in
the real world. To avoid preconceived notions, like "Heads are luckier than
tails," the activities in this section use situations with which students are
not familiar. Each activity allows pupils to collect enough data to predict
the outcome.

Using the Activities

With enough equipment, the entire class can complete one activity and the
results can be compared and combined. Another option is to have small groups
rotate through activity stations which are all on experimental probability
or are on a mixture of topics.

The extra materials needed for these activities include:

. regular dice, three dice per pair of pupils.
. thumbtacks and small paper cups.
. a telephone book.

Pupils may need review on an efficient way to tally, ⅣⅣ groups by fives.

ROLLING SOME SUMS

Jack and Jill are playing a dice game. Each time the dice are rolled, they find the sum of the dots.

Jack predicts a 10 will occur most often.

Jill predicts a 7 will occur most often.

1. Do you (check one blank)

    a. agree with Jack? _____

    b. agree with Jill? _____

    c. think some other sum will occur most often? _____

2. Get two regular dice. Roll them. Find the sum of the dots. Tally the sum in the table to the right.

3. Roll the dice 50 times altogether. Tally.

4. Which sum(s) occurred most often?

   _____

   Which sum(s) occurred least often?

   _____

5. Was Jack's prediction better than Jill's? _____

Sum	Tally	Total
2		
3		
4		
5		
6		
7		
8		
9		
10		
11		
12		

Class Totals

2	3	4	5	6	7	8	9	10	11	12

Rolling Some Sums

Mathematics teaching objectives:

. Develop probability concepts.

. Practice basic addition facts.

. Practice recording skills.

Problem-solving skills pupils might use:

. Collect data needed to solve the problem.

. Use a physical model.

. Make and use a table.

Materials needed:

. Two regular dice per pair of pupils

Comments and suggestions:

. The problem solving skills can be brought out by first presenting
the situation orally and asking questions: "What would you guess?
Can you tell for sure? What is the best guess? How can we find out?
What are the possible sums? How can we keep track of the trials?"
Students will probably suggest "Let's try it and see" (collect data
and use a model), "The sums are 2, 3 ..." (make a list), and "We can
make a table." This discussion provides an excellent opportunity to
verbalize the names of the problem solving skills.

. The use of homemade foam dice will make this a quieter activity.

Answers:

Answers will vary, but seven will usually occur most often and two
and twelve will usually occur least often. This matrix of possible
sums shows why this happens. One such trial of fifty rolls gave the
results shown at the right! Possibly 4 or 5 tally charts could be
compiled into a single chart - seven might then be the most popular
number.

Die A

+	1	2	3	4	5	6
1	2	3	4	5	6	7
2	3	4	5	6	7	8
3	4	5	6	7	8	9
4	5	6	7	8	9	10
5	6	7	8	9	10	11
6	7	8	9	10	11	12

(Die B labels the rows)

Sum	Tally	
2	/	1
3	///	3
4	////	4
5	//// //// /	11
6	//// ////	9
7	////	5
8	////	4
9	//// /	6
10	////	4
11	/	1
12	//	2

DROPPING THUMBTACKS

A dropped thumbtack will land with the point up ⊥ or with the point down ⫮ . If you drop a lot of thumbtacks, do you think (check one):

    a. more will land point up?   _____

    b. more will land point down?   _____

    c. "points up " will be about the
        same as "points down?   _____

Check your prediction by doing this experiment.

1. Get 10 identical thumbtacks and a paper cup.

2. Put the thumbtacks in the cup. Drop them on the floor from a height of 10 centimetres.

3. Count the number landing <u>point up</u> and record in the table below.

4. Repeat the experiment 9 more times. Record.

5. Find the total landing point up and total number dropped.

Experiment	1	2	3	4	5	6	7	8	9	10	Total
Number landing Point Up											
Number dropped	10	10	10	10	10	10	10	10	10	10	

6. Was your prediction accurate?   _____

7. If you repeated the entire experiment, do you think the total number landing point up would be exactly the same?   _____

8. If you repeated the entire experiment, do you think the overall results (most landing point up or point down) would be the same?   _____

<u>Dropping Thumbtacks</u>

Mathematics teaching objectives:

. Develop probability concepts.

. Practice recording skills.

Problem solving skills pupils <u>might</u> use:

. Collect data needed to solve the problem.

. Use a physical model.

. Make and use a table.

. Make predictions after observing patterns.

Materials needed:

. 10 thumbtacks per pair of pupils

. Small cup to hold thumbtacks

Comments and suggestions:

. The problem solving skills can be emphasized by presenting the problem as suggested in the previous activity, or they can be brought out in a summary.

> What was the problem? To decide if more tacks land point up or point down.

> How did we solve it? By trying it with tacks. (Right. We used a model to help collect data.)

> Did the data help us make better predictions? Yes.

. Dropping the thumbtacks on the floor is best so the tacks don't fall, roll, or bounce off the desk or table. Results will be slightly different if the tacks are dropped on a carpet instead of a hard floor.

. If appropriate for your class, pupils could keep a running total and a running percent. After all 10 trials, pupils could find an average number per trial.

Answers:

Although individual trials will show much variation, the overall results will show that most of the tacks will land point up. In previous trials the number of tacks landing point up has been about 65 out of 100 (6 out of 10). This will vary according to tacks used and the type of surface used.

WHICH DO YOU THINK WILL BE LARGER?

1. Two players need three regular dice. Player A gets two dice. Player B gets the other one.

2. Do this activity.

   a. Player A rolls both dice and finds the product of the two numbers.

   b. Player B rolls the one die and multiplies the number by itself.

   c. The winner is the player with the largest answer.

   Who do you think will win more often? _____

3. Do the activity 30 times. Record the results in a table showing three things: A's answer, B's answer, and the winner.

Number	1	2	3	4	5	6	7	8	9	10
A's Answer										
B's Answer										
Winner										

Number	11	12	13	14	15	16	17	18	19	20
A's Answer										
B's Answer										
Winner										

Number	21	22	23	24	25	26	27	28	29	30
A's Answer										
B's Answer										
Winner										

EXTENSION

Do the activity this way:

1. Player A <u>adds</u> the two numbers.
2. Player B <u>adds</u> the number <u>to</u> <u>itself</u>.

Now who do you think will win more often? _____

Which Do You Think Will Be Larger?

Mathematical teaching objectives:

. Develop probability concepts.

. Practice basic multiplication facts.

. Practice recording skills.

Problem solving skills pupils might use:

. Collect data needed to solve the problem.

. Use a physical model.

. Make and use a table.

Materials needed:

. 3 regular dice per pair of pupils

Comments and suggestions:

. The activity can be presented (or summarized) with questions such
as this:

> What is the problem? To decide who will win most often.
>
> How can we solve it? By trying it (use a model and collect data).
>
> How can we keep track of the data? Make a table.
>
> How can we use the data? To make better guesses (make predictions).

. Have pupils do just one trial and record the winner. Pupils can decide
what to do on ties, either re-roll or just count it as a tie.

. Although totals will be close, player B should win more often than
player A. For the extension, the totals should be nearly equal.

. The use of homemade foam dice will make this a quieter activity.

Answers:

Answers will vary. One such trial of 30 rolls gave these results!

Player A	3	10	1	4	30	20	30	18	9	4	5	4	6	12	10
Player B	36	16	1	1	25	25	25	4	1	36	1	1	4	25	4
Winner	B	B	-	A	A	B	A	A	A	B	A	A	A	B	A

Player A	3	10	5	18	12	8	6	15	5	8	6	12	8	4	4
Player B	36	16	1	9	4	4	1	36	16	36	1	36	25	9	1
Winner	B	B	A	A	A	A	A	B	B	B	A	B	B	B	A

B won 13 times.   A won 16 times.   There was one tie.

HELLO, THIS IS I.C. MATH SPEAKING

> I. C. Math's telephone number is 555-3944.
> G. Omtree's number is 555-8788.
>   Add the last two digits of each number.
>   I.C. Math _____   G. Omtree _____

1. If you took 40 telephone numbers and added the last two digits,
    a. what is the smallest sum you could get?   _____
    b. what is the largest sum you could get?   _____
    c. what sum do you think would occur most often? _____

2. Investigate the last question by:

    a. getting a telephone book,

    b. selecting 40 consecutive telephone numbers,

    c. finding the sum of the last two digits,

    d. tallying the results in the table.

3. Was your prediction accurate? _____

4. What would happen if you found the sum of the first two digits of the same 40 telephone numbers?

    _____

    _____

    _____

Sum	Tally	Class Totals
0		
1		
2		
3		
4		
5		
6		
7		
8		
9		
10		
11		
12		
13		
14		
15		
16		
17		
18		

PSM 81

Hello, This Is I.C. Math Speaking

Mathematics teaching objectives:

   . Develop probability concepts.

   . Practice basic addition facts.

   . Practice recording skills.

Problem solving skills pupils <u>might</u> use:

   . Collect data needed to solve the problem.

   . Make and use a table.

Materials needed:

   . Page of a telephone book for each pupil

Comments and suggestions:

   . The suggestions on the previous page (Which Do You Think Will Be Larger?) can be used with this activity.

   . Pupils enjoy getting the page of the telephone book listing their own telephone number (if possible).

   . Like the dice throws in <u>Rolling Some Sums</u>, several sums usually occur more frequently, e.g., 9 (most often), then 8 or 10, then 7 or 11.

   . Using the first two digits of the same 40 telephone numbers will show greatly distorted results since the prefix numbers usually are similar.

Answers:

   Answers will vary. A trial of 40 telephone numbers gave these results.

Combined class results usually show that those sums in the middle do occur more often.

Number	Tallies	Count
0	/	1
1	/	1
2		0
3	/	1
4	///	3
5		0
6	/	1
7	/HN /	6
8	///	3
9	////	4
10	///	3
11	/HN //	7
12	///	3
13	/	1
14	///	3
15	/	1
16	///	3
17		0
18		0

Grade 5

VIII.   ESTIMATION WITH CALCULATORS

VIII. ESTIMATION WITH CALCULATORS

Estimation should be an important part of the mathematics curriculum. Some authorities say that 75% of the adult non-occupational uses of math involve mental arithmetic and estimations. Frequently it is _not_ necessary to have an exact answer, amount, or measurement. Pupils need many exposures to the estimation process. Some are reluctant to try. They would rather work a somewhat complicated problem in order to get "the" exact answer. The process needs to be taught (and practiced) using such individual skills as computing with single digits, rounding, and performing operations with powers and multiples of ten.

Using the Activities

The five activities in this section use whole numbers and money amounts. A knowledge of how to use a calculator is assumed. If your pupils have not used them before, several readiness activities with calculators would be appropriate. Several calculator books now on the market can provide readiness activities. One, in particular, is CALCULATOR ACTIVITIES FOR THE CLASSROOM by Immerzeel and Ockenga, available from Creative Publications.

Special emphasis is placed on first making estimates and then using a calculator to check the estimates. It is better if all estimates are made before the calculators are passed out. For many of these activities, pupils can work in pairs and share a single calculator.

If you do _not_ have calculators available and/or your pupils have mastered the basic skills of addition, subtraction, multiplication with 2-digit multipliers and division with 2-digit divisors, these activities can be done _without_ calculators.

For each of the activities done, be sure to mention the problem solving skills that are appropriate. The activity Grocery Shopping can be introduced by describing a person at a grocery store with only a five dollar bill. Estimating the total cost as groceries are selected helps solve the problem of staying within the limit. Another way to motivate estimation is to ask students how they know a calculator gives the right answer. Discussion can bring out that people can push the wrong buttons and use the calculator incorrectly. Weak batteries can also cause incorrect answers. Estimating answers can help in catching errors. Emphasize that a good way to check is to solve a problem using different procedures (estimating and by calculator). Students can share hints and strategies for the last three activities. Some might identify the skill, guess, check and refine.

SPENDLESS MARKET (Several years ago!)

Peanut Butter $1.98

Margarine 38¢ lb.

Syrup $1.49

Cottage Cheese 44¢ pt.

Grape Jelly 99¢

Top Sirloin $2.29 lb.

Tomato Juice 48¢

Bread 39¢

Coffee $2.69 lb.

Ground Beef 88¢ lb.

Bananas 29¢ lb.

Potato Chips 69¢

Read <u>all</u> directions first.

- . Estimate the total cost of the items on each grocery list. Record in the estimate part of the table on the next page.

- . After you've made all of your estimates, get a calculator and determine the actual costs. Record your answers in the table.

- . How does each estimate compare with the actual cost? Indicate this by drawing a smiling face or a frowning face (or possibly a straight face).

Grocery Shopping

Mathematics teaching objectives:
- . Estimate.
- . Compute with money amounts.
- . Use a calculator.

Problem-solving skills pupils might use:
- . Make reasonable estimates as answers.
- . Solve a problem by using different procedures.

Materials needed:
- . Calculator

Comments and suggestions:
- . An overhead transparency of the prices would eliminate the need for each pupil to have a copy of the first page.
- . Each pupil should first make an estimate for each of the 12 problems. This may serve as an activity for one day with the actual checking with a calculator taking place the next day. A time limit may encourage pupils to really estimate rather than computing actual amounts.
- . Observe and discuss how pupils work problems 3, 5, 7, 10 and 11. Their estimates may be reasonable. But the calculator answer may be wrong if they enter the data in the following order -- $\boxed{2}\,\boxed{\times}\,\boxed{.}\,\boxed{2}\,\boxed{9}\,\boxed{+}\,\boxed{2}\,\boxed{\times}\,\boxed{.}\,\boxed{6}\,\boxed{9}\,\boxed{=}$. It is better to use an application of the distributive property and enter the data as $\boxed{.}\,\boxed{2}\,\boxed{9}\,\boxed{+}\,\boxed{.}\,\boxed{6}\,\boxed{9}\,\boxed{=}\,\boxed{\times}\,\boxed{2}\,\boxed{=}$.
- . A discussion about how pupils feel about their estimates is important. Emphasize that some estimates are closer than others but seldom is any estimate wrong--unless it is extremely far off.
- . Alert pupils to proper way of entering 88¢ into the calculator. Some will want to just touch $\boxed{8}\,\boxed{8}$ instead of $\boxed{.}\,\boxed{8}\,\boxed{8}$.

Answers:

1.	$2.75	7.	$4.62
2.	$16.03	8.	$4.22
3.	$1.96	9.	$7.45
4.	$13.45	10.	$8.54
5.	$4.14	11.	$3.10
6.	$3.84	12.	$12.99

Grocery Shopping (cont.)

	Estimate	Answer	🙂 or 🙁
1. 1 peanut butter 1 margarine 1 bread			
2. 7 lb. top sirloin			
3. 2 lb. bananas 2 packages potato chips			
4. 5 lb. coffee			
5. 3 grape jelly 3 bread			
6. 8 tomato juice			
7. 6 margarine 6 bread			
8. 1 of each - peanut butter, jelly, margarine, tomato juice, bread			
9. 5 syrup			
10. 2 lb. top sirloin 2 peanut butter			
11. 2 lb. each of - margarine, bananas, ground beef			
12. What is the total cost of everything listed in the picture?			

BEST GUESS

Get a calculator and play this game with another person.

Rules

. Each player estimates the answer to Problem 1. Be sure to write down your estimates. You should decide how much time to allow for each estimate.

. Now use a calculator to determine the correct answer.

. The player whose estimate is closest gets one point.

. Continue with the rest of the problems using the same rules.

1. 972 - 399 =

2. 59 + 74 + 308 + 183 =

3. 19 x 21 =

4. 29 x 31 =

5. 15 ⟌6345

6. $108.19 - $59.99 =

7. 1075 ÷ 25 =

8. (78 + 148) ÷ 2 =

9. (47 + 39) x 9 =

10. 5438 - 499 =

11. 8 x 12 x 9 x 10 x 15 x 7 =

12. 11 x 11 x 11 =

13. 57 + 93 + 84 - 55 - 95 =

14. The sum of all the numbers from 1 to 20

<u>Best Guess</u>

Mathematics teaching objectives:

. Estimate.

. Use a calculator.

Problem solving skill pupils <u>might</u> use:

. Make reasonable estimates as answers.

Materials needed:

. Calculator

Comments and suggestions:

. A discussion of skills used in estimation, that is, rounding, computing with multiples of 10, etc., makes a good lead-in for this activity.

. Each pair of pupils needs a calculator. After estimates are made by both players, they use the calculator to find the exact answer. The calculator can also be used to determine how far off each estimate is.

. In keeping with the idea that no estimate is wrong, a variation of the scoring would be to award two points for the closer estimate and one point for the other estimate.

. Pupils enjoy creating problems which they think will be too difficult for others. Have them create ten or twenty problems to be used with this game.

Answers:

1.	573	8.	113
2.	624	9.	774
3.	399	10.	4939
4.	899	11.	907,200
5.	423	12.	1331
6.	$48.20	13.	84
7.	43		

NUMBER HUNT

Example 1:

Which two numbers at the right will have a sum closest to 85?

Eric found that 43 + 39 = 82.

Can you find any that are closer?

Example 2:

Which two numbers at the right will have a sum closest to 100?

This time, Eric used 43 + 52 = 95.

Sonja found a closer pair.

Which numbers do you think Sonja used?

1. Which two numbers at the right will have a sum close to

   a. 125 ?
   b. 140 ?
   c. 150 ?
   d. 175 ?

2. Which three of the numbers will have a sum close to

   a. 150 ?
   b. 200 ?
   c. 250 ?
   d. 300 ?

39

43

52

65

74

79

81

84

90

95

EXTENSION

Which 9 numbers will have a sum closest to 650 ?

<u>Number Hunt</u>

Mathematics teaching objectives:
- . Estimate.
- . Use a calculator.

Problem-solving skills pupils <u>might</u> use:
- . Make reasonable estimates as answers.
- . Guess and check.
- . Study the solution process.
- . Recognize limits and eliminate possibilities.

Materials needed:
- . Calculator

Comments and suggestions:
- . As a whole class, go over the examples emphasizing the estimation process. Pupils will not find two numbers with a sum closer to 85 than 43 and 39. But in the second example, 65 + 39 is closer to 100 than is 43 + 52.
- . For the problems, pupils have to decide what is meant by "close." In most cases it is possible to get within five of the desired sum.
- . Encourage pupils to study what they have done in problem 1. Some may use their answers to help make the estimates in problem 2.

Answers:

Some possible answers are:

1. a. 43 + 81 = 124
        39 + 84 = 123

    b. 43 + 95 = 138
        65 + 74 = 139
        52 + 90 = 142

    c. 65 + 84 = 149
        52 + 95 = 147

    d. 81 + 95 = 176
        84 + 90 = 174
        79 + 95 = 174

2. a. 39 + 43 + 65 = 147

    b. 43 + 65 + 95 = 203

    c. 74 + 79 + 95 = 248

    d. 84 + 90 + 95 = 269

For the extension, all ten numbers have a sum of 702. The nine numbers with a sum closest to 700 will be the nine numbers obtained by eliminating 39.

-234-

A NUMBER TIMES ITSELF

Use a calculator. Fill in the table.

14 x 14	
24 x 24	
34 x 34	
44 x 44	
54 x 54	
64 x 64	
74 x 74	
84 x 84	
94 x 94	

Now solve these problems. <u>In each case the number in the two</u>
<u>boxes must be the same</u>.

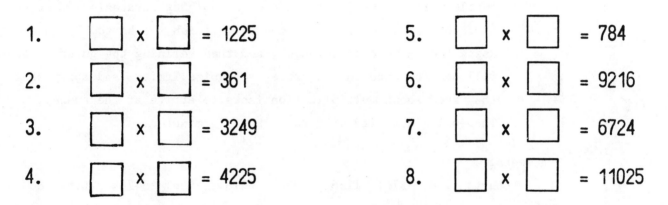

1. ☐ x ☐ = 1225   5. ☐ x ☐ = 784

2. ☐ x ☐ = 361   6. ☐ x ☐ = 9216

3. ☐ x ☐ = 3249   7. ☐ x ☐ = 6724

4. ☐ x ☐ = 4225   8. ☐ x ☐ = 11025

EXTENSION

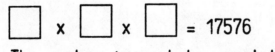

☐ x ☐ x ☐ = 17576

The number in each box must be the same.

A <u>Number</u> <u>Times</u> <u>Itself</u>

Mathematics teaching objectives:

    . Estimate.

    . Use a calculator.

Problem solving skills pupils <u>might</u> use:

    . Make and use a table.

    . Recognize limits and eliminate possibilities.

    . Make reasonable estimates as answers.

    . Guess and check.

Materials needed:

    . Calculator

Comments and suggestions:

    . The chart at the top could be filled in as a whole-group activity. This would allow you to determine if all pupils were correctly using the calculator. There is no intended pattern in the products except that all end in a 6.

    . The remainder of the sheet should be an individual (or pair) task. Observe the pupils using the guess, check and refine strategy. Perhaps a hint or two about using the chart will help.

    . Pupils will use two distinct problem solving strategies after the table has been completed. Some will finish the bottom by guessing, using the calculator to check, and then refining the guess. Others will use the data in the table, recognize limits, and apply what they know about multiplication facts to arrive at the answer.

    . The entire activity will take about 25 minutes.

Answers:

    Chart: 196;   576;   1156;   1936;   2916;   4096;   5476;   7056;   8836

1.	35	5.	28
2.	19	6.	96
3.	57	7.	82
4.	65	8.	105

    Extension: 26

MISSING PARTS

The numbers in each column are to be used to complete the problems involving that operation. For example, the numbers in the addition column are to be used to complete the addition problems. An extra number is contained in each column.

Addition	Subtraction	Multiplication
999	149	2337
127	78	49
98	487	25
569	829	22
1266	646	234
164	201	19
437	991	505

1.
$$\frac{\bigcirc + \bigcirc}{1006}$$

2.
$$+ \quad 267$$
$$\bigcirc$$

3.
$$\frac{\bigcirc + \bigcirc}{262}$$

4.
$$\frac{\bigcirc - \bigcirc}{751}$$

5.
$$\bigcirc - 52$$
$$\bigcirc$$

6.
$$\bigcirc - 504$$
$$\bigcirc$$

7.
$$\frac{\bigcirc \times \bigcirc}{1078}$$

8.
$$123$$
$$\times \bigcirc$$
$$\bigcirc$$

9.
$$\frac{\bigcirc \times \bigcirc}{12625}$$

-237-

PSM 81

Missing Parts

Mathematics teaching objectives:

- Estimate.
- Use a calculator.
- Use inverse operations (try 1006 - 999 to see if it will work in problem 1, an addition problem).

Problem-solving skills pupils <u>might</u> use:

- Make reasonable estimates as answers.
- Guess and check.
- Recognize limits and eliminate possibilities.
- Apply what you know about basic facts.

Materials needed:

- Calculator

Comments and suggestions:

- Two distinct problem-solving strategies are possible. Some pupils will use a guess and check approach. Others will analyze the data, recognize limits, and apply what they know about basic facts.
- Remind pupils that each column contains an extra number. They can cross off each number as it is used.
- Pupils might enjoy making up their own problems to share with others.

Answers:

1. 569 + 437 = 1006
2. 999 + 267 = 1266
3. 164 + 98 = 262
4. 829 - 78 = 751
5. 201 - 52 = 149

6. 991 - 504 = 487
7. 49 x 22 = 1078
8. 123 x 19 = 2337
9. 505 x 25 = 12,625

Extension:

To challenge pupils, use these division problems.

Possible answers are: 80,000;   61;   735;   6144;   24;   549;   625

Grade 5

IX. CHALLENGES

IX. CHALLENGE PROBLEMS

The activities in the Getting Started section were very directed and pupils were encouraged to use (although not completely restricted to) one problem-solving skill at a time. The challenge problems in this section leave the choice of the problem-solving method up to the pupil. The intention is to allow for and encourage individual differences, creativity and cooperation.

Let's look at how one challenge problem, "Machine Hook-Ups," can be used in the classroom. As you read the example, notice how the teacher does not structure or direct the methods pupils use, but that the teacher does have these important functions:

- . to help students understand the problem.
- . to listen if pupils want to discuss their strategies.
- . to praise and encourage pupils in their attempts, successful or not.
- . to facilitate discussion of the problem and sharing of the strategies.
- . to give hints or ask questions, if necessary.
- . to summarize or emphasize methods of solution after pupils have solved the problem.

As we look into Mr. Lane's classroom, the page, "Machine Hook-Ups" has been distributed. Pupils are used to seeing a different challenge each week and they know they will be given some classtime to work on the problem.

Mr. L: Here is the challenge problem for the week. I'll let you look at the problem, then we'll discuss it to be sure we all understand it. (Waits.) Merlin, you look puzzled.

Merlin: I'm not sure what to do at the second box. Do I multiply 8 by 4 or 14 by 4?

Mr. L: The result from one box is put into the next box. You got 14 by adding 6 to 8, so 14 is what you multiply by 4. Using that idea, why don't you all try to find the last output when 8 is put into the top box? (Walks around the room to look at work.) Some of your answers aren't agreeing. I see 27's, 23's and 28's. Will you please check each others work and see if you can agree? You all agree on 23? Good. If you think you know how the machine works, go ahead and fill out the rest of the table. If you're still having trouble with it, raise your hand. (Students compare outputs as they go and Mr. Lane walks around the room answering questions and checking computation. Puzzled expressions occur as pupils see they have to figure out what has to be put into the machine to get 39 out.)

Pearl: Mr. Lane, I don't know how to work the last two.

Mr. L: It wouldn't be a challenge if you knew right away, would it? What kinds of things can you try?

Pearl: (Looking up at the posted list of problem-solving skills.) I'll try guessing.

Mr. L: Good idea. (In a quiet, side conversation with Peggie.) Peggie, you have found that 16 had to be put in the machine to get 39 out. How did you do that?

Peggie: I saw that 15 gave 37 so I tried numbers bigger than 15. 17 was too big, but 16 worked.

Mr. L : That was a good method. You saw a pattern in the table and then guessed and checked. I'll let you work on the last one. (More individual discussions and work goes on.) Most of you are done with the table. Would anyone like to share their methods of solving the last two problems? (Several describe what they did and Mr. Lane summarizes as necessary.). Later you will get a chance to make your own input-output machine. You all put a lot of effort into this problem. I'm glad to see so many good problem-solvers!

The above approach to challenge problems also gives opportunities for practicing the following problem-solving skills:

. State the problem in your own words.

. Clarify the problem through careful reading and by asking questions.

. Share data and the results with other interested persons.

. Listen to persons who have relevant knowledge and experiences to share.

. Study the solution process.

. Invent new problems by varying an old one.

> # IN GENERAL, THE TEACHER MUST BE AN ACTIVE, ENTHUSIASTIC SUPPORTER OF PROBLEM SOLVING.

Using the Activities

Sixteen challenge problems are provided. Some teachers give them as a "challenge of the week" or as a Friday activity. On the day the challenge is given out, time should be spent on getting acquainted with the problem. On following days, a few minutes can be devoted to pupil progress reports. If there is little sign of progress, you can provide some direction by asking a key question or suggesting a different strategy. At appropriate times, the activity can be summarized by a class discussion of strategies used and some problem extensions.

Key problem-solving strategies that pupils have used in solving the problems are given in the comments for each problem. Your pupils might have additional ways of solving the problems.

One Plan For Using A Challenge   (over a period of 1 or 2 weeks)

First day -

. Give out the challenge. (Possibly near the end of the period.)

. Let pupils read written directions and possibly discuss with
  a classmate.

. Clarify any vocabulary which seems to be causing difficulty. Ask
  a few probing questions to see if they have enough understanding
  to get started.

. Remind them that during a later math class, time will be used to
  look at the problem again.

Later in the week -

. Have pupils share their ideas.

. Identify the problem-solving skills suggested by these ideas.

. Conduct a brainstorming session if pupils do not seem to know
  how to get started.

. Suggest alternative strategies they might try.

. Give an extension to those pupils who have completed the challenge.

On a subsequent day -

. Allow some class time for individuals (or small groups) to work on
  the challenge. Observe and encourage pupils in their attempts.

. Try a strategy along with the pupils (if pupils seem to have given
  up.)

Last day -

. Conduct a session where pupils can present the unsuccessful as well
  as the successful strategies they used.

. Possibly practice a problem-solving skill that is giving pupils difficulty; e.g., recording attempts, making a systematic list, or checking
  solutions.

Key problem-solving strategies that pupils have used in solving the
problems are given in the comments for each problem. Your pupils might
have additional ways of solving the problems.

A challenge problem for the teacher: Keep the quick problem solver
from telling answers to classmates.

HEXAGON PUZZLE

Place the numbers 1-19 in the circles so the sum of the
rows with three circles (o-o-o) is 23 and the sum of the rows
with five circles (o-o-o-o-o) is 40. Use each number only once.

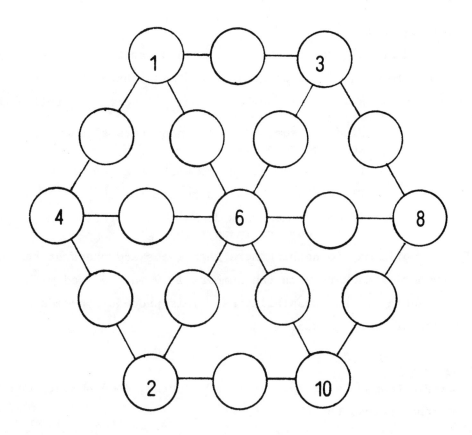

X 2 3 A 5 6 7 8 9 10 11

12   13   14   15   16   17   18   19

Hexagon Puzzle

Problem-solving skills pupils <u>might</u> use:

. Guess and check.

. Break a problem into manageable parts.

. Eliminate possibilities.

Comments and suggestions:

. Some students will break the problem into parts, finding the numbers in the outer ring first using addition and subtraction. After the outer ring is solved, six numbers remain. Pupils may follow these steps:

    a. Find the sum of the given numbers already in a 5-circle row.

    b. Find two numbers to complete the sum of 40.

    c. Place the numbers and repeat for the other rows.

    d. If one or more sum doesn't work, refine the choice of the numbers in part (b).

. Other pupils may do nothing more than guess and check at the placement of numbers. Markers with the numbers 1-19 may be used with guessing and checking to try possibilities. This procedure avoids a lot of erasing of wrong guesses.

Answer:

Two distinct answers are possible. Of course, some numbers could also be switched across the 6.

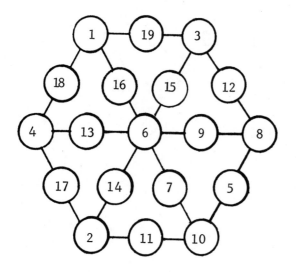 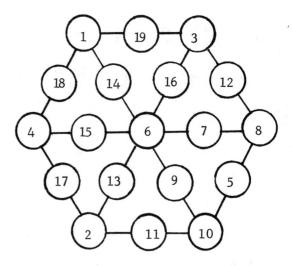

SQUARE PATTERNS

John found a box with 1024 small square tiles in it. He wonders how large a square he could make by putting small square tiles together.

1. Help John decide by making smaller squares. Look for a pattern. Predict the largest square he can make.

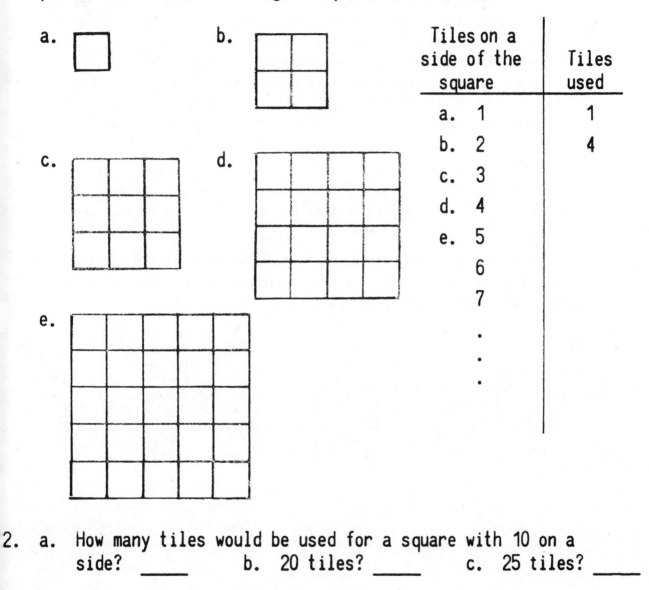

Tiles on a side of the square	Tiles used
a. 1	1
b. 2	4
c. 3	
d. 4	
e. 5	
6	
7	
.	
.	
.	

2. a. How many tiles would be used for a square with 10 on a side? ____   b. 20 tiles? ____   c. 25 tiles? ____

3. What is the largest square John could make? _____

Square Patterns

Problem-solving skills pupils <u>might</u> use:

  . Look for and use a pattern.

  . Make and use an organized table.

  . Guess and check.

Comments and suggestions:

  . You might want to give this problem to students <u>orally</u> without any hints on how to solve it. After they understand the problem, ask them how they might solve it. Some might ask for 1024 tiles to try it (use a model). Others might try to solve it on grid paper (make a diagram). Both strategies also use guess, check, and refine. After discussion, the dittoed page can be given as a third way to solve the problem. It shows how to generalize from a series of examples or similar problems.

  . Pupils may recognize the pattern of multiplying the number of tiles on a side by itself to find the total number of tiles. Question 2 should motivate more pupils to look for a general rule so they won't have to use tiles or grid paper.

  . To answer the final question, pupils might be asked what numbers in the units place would give a 4 in the units place of the product when the number is multiplied by itself.

Answers:

Tiles on a side	1	2	3	4	5	6	7	...	10	...	20	...	25	...	32
Total Tiles	1	4	9	16	25	36	49		100		400		625		1024

STAR WARS

1. Put different numbers in each circle. The sum of the four numbers along any line should be 50.

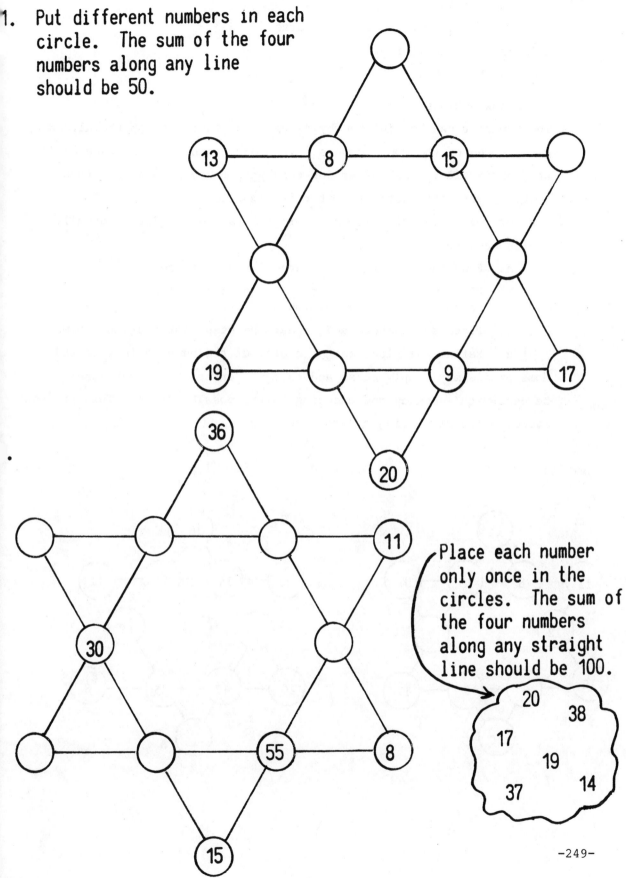

Place each number only once in the circles. The sum of the four numbers along any straight line should be 100.

20 38
17
19
37 14

PSM 81

Star Wars

Problem-solving skills pupils might use:

. Break a problem into manageable parts.

. Eliminate possibilities.

. Guess and check.

Comments and suggestions:

. The top star can be finished by recognizing that the top row already
gives three of the four numbers. The fourth can be found by addition
and subtraction. The answer for this part makes it possible to do
another part, etc. until the star is complete.

. Part of the bottom star is worked in the same way. But eventually
pupils need to:

    a. Find the sum of the numbers already entered in a row.

    b. Find two numbers from the list of possible numbers
       to complete the sum of 100.

    c. Place the numbers and repeat the steps for the other rows.

    d. Refine the placement or choice of the numbers in part (b).

. Some pupils may simply guess and check. Markers with the numbers may
be used with guessing and checking to try possibilities. This procedure
avoids a lot of erasing of wrong guesses.

Answers:

1.

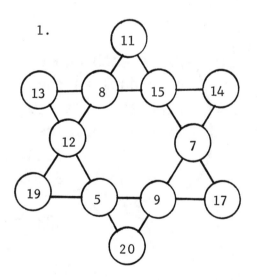

2.

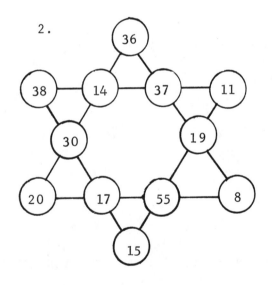

A LOGICAL CHOICE

1. Larry, Curly, and Moe left the theater together. Each one was wearing the jacket and the shoes of someone else.

   Curly was wearing Moe's jacket and Larry's shoes.

   Whose jacket and whose shoes was each one wearing?

2. a. A yuk is more than an ugh.
   b. A bah is not the least of the whole group.
   c. A yuk is not the greatest.
   d. Only one thing is less than an ick.
   e. A glok is more than a bah.
   f. More than one thing is greater than a bah.

   Put the yuk, ugh, bah, ick, and glok in order from largest to smallest.

PSM 81

<u>A Logical Choice</u>

Problem-solving skills pupils <u>might</u> use:

. Break a problem into maneageable parts.

. Make and use a table.

. Eliminate possibilities using contradictions.

. Guess and check.

Comments and suggestions:

. Logic problems lend themselves to using a table. List one variable across the top. Use of yes and no, or X and 0, or any convenient notation permits the problem solver to use each clue to find the answer.

	Larry	Curly	Moe
Jacket			
Shoes			

. Students can solve the second problem by guessing and checking. Some may write the names on slips of paper and move them around to check various arrangements. A table could also be used to solve this problem.

Squares are X'd out if it is not a possible position. The bah can't be least. The ick is second, so it is written in and the rest of it's row and column are X'd out. (No one else can be second.) etc.

	least ——————→ most				
yuk		X			
ugh		X			
bah	X	X			
ick	X	ick	X	X	X
glok		X			

Answers:

1.
	Larry	Curly	Moe
Jacket		Moe's	
Shoes		Larry's	

(Given)

	Larry	Curly	Moe
Jacket	Curly's	Moe's	Larry's
Shoes	Moe's	Larry's	Curly's

Larry cannot wear his own jacket so he must wear Curly's.

Moe cannot wear his won shoes so he must wear Curly's

2.
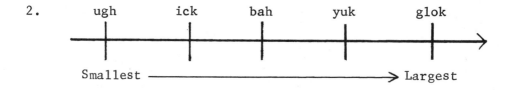

ugh   ick   bah   yuk   glok

Smallest ————————————→ Largest

-252-

MAGIC SQUARES

This magic square uses the numbers
1 to 9. Each row, column, and
long diagonal adds to the magic
sum of 15.

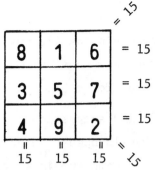

1. Complete these magic squares. Write the magic sum for each.

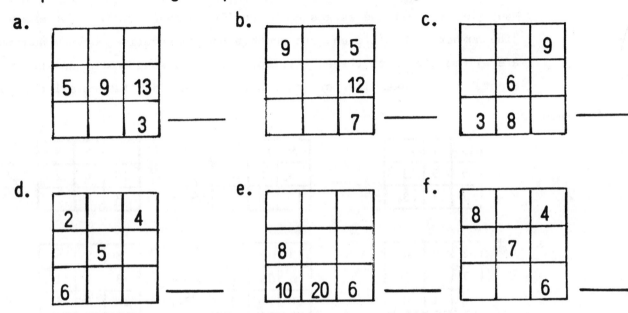

2. Many patterns can be found inside magic squares. For
    example; add the corner numbers, divide by four, and
    you will get the middle number. Study the magic squares
    above. Write the patterns you find.

3. Use your patterns to complete these magic squares. Write the
    magic sum for each.

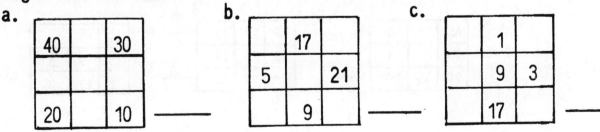

Magic Squares

Problem-solving skills pupils <u>might</u> use:

- Break a problem into manageable parts.
- Look for a pattern and make predictions.

Comments and suggestions:

- Pupils need to understand that all rows, columns, and diagonals give the same sum in a magic square. For each problem enough information is given to determine this sum.
- Many patterns will be suggested by pupils to answer (2). For example, "The numbers are all even, the numbers are consecutive, or the numbers are all multiples of three." Accept any reasonable answer and try to create counter-examples for those patterns that don't work all the time.

Answers:

1.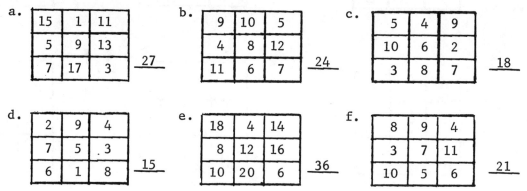

   a.

15	1	11
5	9	13
7	17	3

   27

   b.

9	10	5
4	8	12
11	6	7

   24

   c.

5	4	9
10	6	2
3	8	7

   18

   d.

2	9	4
7	5	3
6	1	8

   15

   e.

18	4	14
8	12	16
10	20	6

   36

   f.

8	9	4
3	7	11
10	5	6

   21

2. Some patterns that work for all the magic squares:

   - Add the corner numbers and divide by 4 and you get the middle number.
   - Add the middle number of each side and divide by 4 and you get the middle number.
   - Add the first two numbers in a diagonal and divide by 2 and you get the opposite corner number.
   - The magic sum is a multiple of 3.

3. a.

40	5	30
15	25	35
20	45	10

   75

   b.

15	17	7
5	13	21
19	9	11

   39

   c.

10	1	16
15	9	3
2	17	8

   27

LOTS OF ZEROS

1. Solve this addition problem:

$$
\begin{array}{r}
111 \\
333 \\
555 \\
777 \\
+ \ 999 \\
\hline
\end{array}
$$

2. Change 12 digits in the
   problem to zeros and get
   a sum of 20.

   Record ────────────→

3. Change 10 digits to
   zeros. Get a sum of 100.

   Record ────────────→

4. Change 8 digits to
   zeros and get a sum
   of 1,111.

   Record ────────────→

5. Change 9 digits to
   zeros and get a sum
   of 1,111.

   Record ────────────→

PSM 81

<u>Lots of Zeros</u>

Problem-solving skills pupils <u>might</u> use:

. Work backwards.

. Guess and check.

. Satisfy one condition at a time.

. Recognize limits and eliminate possibilities.

Comments and suggestions:

. You may need to suggest that both 11 and 011, for example, stand for eleven.

. Pupils will likely start the problems by using guess, check, and refine.

. If they seem to be struggling too much you might suggest that satisfying one condition at a time and then making adjustments might be worth trying. For example, in problem 2 the 1, 3, 7 and 9 does have a sum of twenty as required, but the number of zeros you need to use is 11 rather than 10. Now adjustments need to be made.

. Another suggestion that might be made - start adding from left to right rather than from right to left.

Answers:

1.   2775

2.*	010	3.*	011	4.*	101	5.*	111
	003		030		300		030
	000		050		550		000
	007		000		070		070
+	000	+	009	+	090	+	900
	020		100		1,111		1,111

*Other solutions are possible.

Extension:

Try to get 1,111 as a sum by changing 10 digits to zero.

	111
	300
one answer	000
	700
+	000
	1,111

STRANGE HAPPENINGS

	0	1	2	3	4	5	6	7	8	9
10	⑪	12	⑬	14	15	16	17	18	19
20	21	22	23	24	25	㉖	27	㉘	29
30	㉛	32	㉝	34	35	36	37	38	39
40	41	42	43	44	45	㊻	47	㊽	49
50	51	52	53	54	55	56	57	58	59
60	61	62	63	㊳	65	㊺	67	68	69
70	71	72	73	74	75	76	77	78	79
80	81	82	83	㊸	85	㊻	87	88	89
90	91	92	93	94	95	96	97	98	99

1. For each 3 by 3 (those using circles)

   a. Multiply the two pairs of numbers connected by a line.
   b. Record in the table below.
   c. Subtract the smaller product from the larger product. Record.
   d. Repeat these steps with a 3 by 3 of your own.

	Product	Product	Difference
1			
2			
3			
Your own			

What did you discover?

_____

_____

_____

_____

2. Try different sizes. Record the difference between the products. A 2 by 2 is shown using squares.

Size	2 by 2	3 by 3	4 by 4	5 by 5
Difference				

3. Predict the result of a 6 by 6. _____ Check your prediction.

4. Predict the result of a 10 by 10. _____ Check your prediction.

Strange Happenings

Problem-solving skills pupils <u>might</u> use:

. Look for and use patterns.
. Make a systematic list.
. Guess and check.

Comments and suggestions:

. Use of a calculator frees pupils to concentrate on the patterns developed in this activity.

. These X's have both dimensions the same, i.e. 2 by 2, 3 by 3, etc. An interesting extension is to investigate a 3 by 3, a 3 by 4, a 3 by 5, a 3 by 6, etc. or perhaps a 4 by 4, a 4 by 5, a 4 by 6, etc.

Answers:

1.

	Product	Product	Difference
1	363	403	40
2	1248	1288	40
3	5504	5544	40
Your own			

The difference between the products of 3 by 3 X's is 40.

2.

Size	2 by 2	3 by 3	4 by 4	5 by 5
Difference	10	40	90	160

3. The 6 by 6 difference will be 250. (Pupils may see a pattern in the differences.

10 ⟍ 40 ⟍ 90 ⟍ 160 ⟍ ?
   30   50   70   90

So the next term must be 250 to have 250 - 160 = 90.)

4. The 10 by 10 difference will be 810.

DUCKS AND COWS

1. Farmer McDonald raises ducks and cows.

   The animals have a total of
   9 heads and 26 feet.

   How many ducks and how many
   cows does Mr. McDonald have?

Try to solve these "ducks and cows" puzzles. If a puzzle is
not possible, explain why.

2. 9 heads and 20 feet

3. 10 heads and 24 feet

4. 8 heads and 18 feet

5. 9 heads and 50 feet

6. 6 heads and 17 feet

7. 10 heads and 18 feet.

EXTENSION

Farmer McDonald raises ducks and cows. He is out standing
in his field and sees some of each kind of animal. Altogether he sees 24 feet (not including his own!). How many
ducks and how many cows does he see? Show all possible
answers.

Ducks and Cows

Problem-solving skills pupils <u>might</u> use:

 . Guess and check.

 . Make a systematic list.

 . Make a drawing.

Comments and suggestions:

 . Pupils may guess at the number of cows and ducks, check their guess and refine according to whether the number of feet was too large or too small.

 . Other pupils may make a systematic list and show possibilities for 9 cows, 0 ducks; 8 cows, 1 duck; 7 cows, 2 ducks; etc. until the correct solution is found.

 . Other pupils may make a drawing of the nine heads. The feet are passed out two at a time until each head has at least two feet. Then the remaining feet are passed out two at a time until none remain. The ducks and cows can then be counted.

Answers:

1. 4 cows, 5 ducks

2. 1 cow, 8 ducks

3. 2 cows, 8 ducks

4. 1 cow, 7 ducks

5. Not possible. 36 is the most feet you could have.

6. Not possible. The number of feet must be even.

7. Not possible. 20 is the least feet you could have.

Extension: There are 5 possible answers.

Cows	Feet	Ducks	Feet	Total
5	20	2	4	24
4	16	4	8	24
3	12	6	12	24
2	8	8	16	24
1	4	10	20	24

MACHINE HOOK-UPS

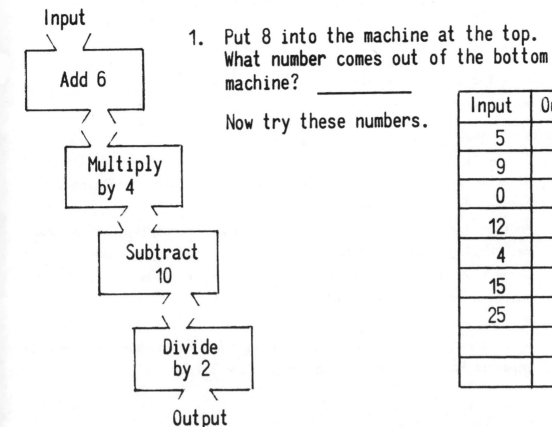

Input

Add 6

Multiply
by 4

Subtract
10

Divide
by 2

Output

1. Put 8 into the machine at the top.
   What number comes out of the bottom
   machine? _____

   Now try these numbers.

Input	Output
5	
9	
0	
12	
4	
15	
25	
	39
	47

2. Use the drawing at the right to
   "build" your own machine hook-ups.
   Use the numbers below in your
   machine.

   Does your table have fractions?
   If so, try to change your machine
   so that this
   will not
   occur.

Input	Output
0	
5	
10	
	21

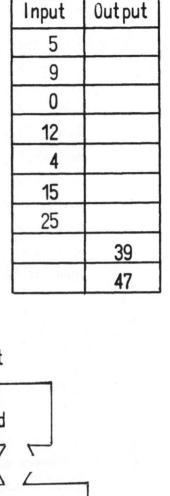

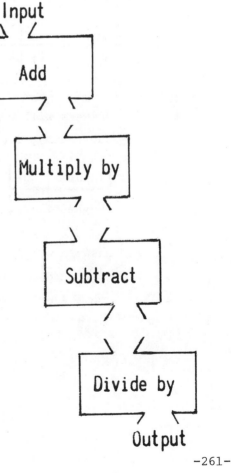

Input

Add

Multiply by

Subtract

Divide by

Output

EXTENSION

Build a machine with four hook-ups in
which an input of 5 will give an
output of 11.

PSM 81

Machine Hook-Ups

Problem-solving skills pupils <u>might</u> use:
- . Guess and check.
- . Look for and use patterns.
- . Work backwards.

Comments and suggestions:
- . The first seven blanks in the table can be filled in by following directions. Pupils may use guess and check for the remaining two blanks.
- . Another approach involves studying the table carefully, maybe even rearranging the entries to put them in order by input numbers. These show that the inputs for 39 and 47 are between 15 and 25, with one input very close to 15.
- . A third approach would be to work backwards and use the inverse operations of those given. If this is done, pupils should check the input to be sure it gives the appropriate output.

Answers:

1. Input 8 ⟶ output 23

Input	5	9	0	12	4	15	25	16	20
Output	17	25	7	31	15	37	57	39	47

2. Answers will vary. One possible hook-up is: Add 5, multiply by 6, subtract 3, divide by 3.   This gives

Input	0	5	10	6
Output	9	19	29	21

3. Answers will vary. Two possible hook-ups are:
    a. Add 6, multiply by 5, subtract 11, divide by 4.
    b. Add 5, multiply by 11, subtract 11, divide by 9.

A PIG PEN

ally and Sam Jones own a farm. One day they were out standing
their field wondering how to pen up the pigs. A big pen can
divided into small pens using exactly 3 straight fences.
how where Sally and Sam should build the 3 fences to put each
ig in a separate pen. The ends of each fence must reach the
ides of the square.

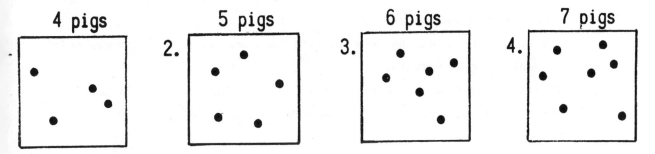

uppose Sally and Sam used 4 straight fences. What numbers of
igs could they pen? Draw the fences and place a dot for each
ig.

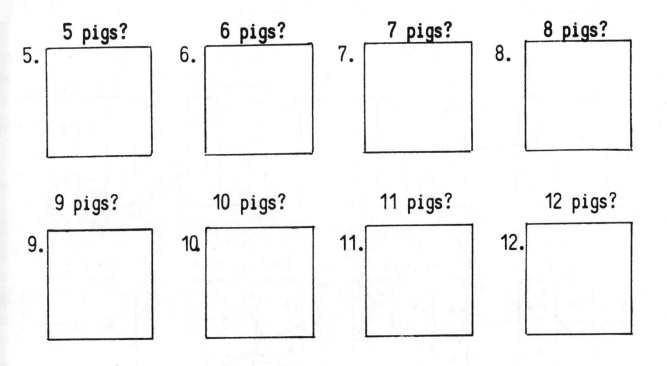

A Pig Pen

Problem-solving skills pupils <u>might</u> use:
- Make and use a drawing.
- Guess and check.
- Look for patterns.

Comments and suggestions:
- Straws or pieces of uncooked spaghetti could serve as models for the fences. This procedure avoids a lot of erasing of errors. The fences should end on the squares and there can be empty pens.
- Several problems can be done in more than one way. The answers below are only sample answers.
- Pupils may pick up patterns in how parallel fences create a certain number of pens opposed to intersecting fences.

Answers:

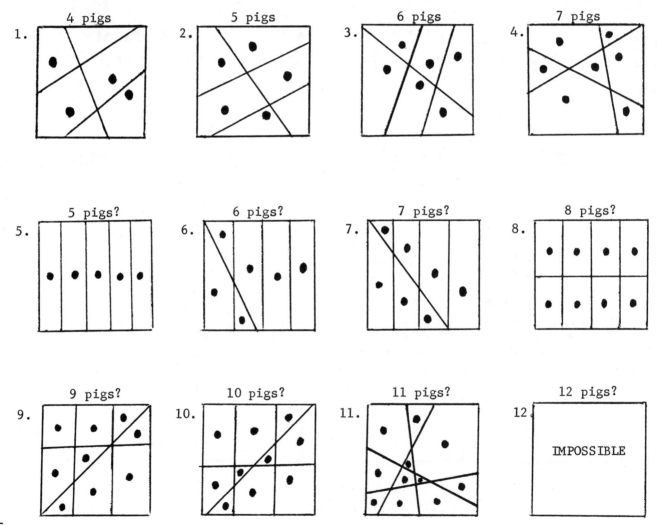

DECIMAL FRAMES

GET 3 markers

Show all the different decimals you can make by placing
all 3 markers on this frame. The example below shows 1.02.
Record the answers on the frames below.

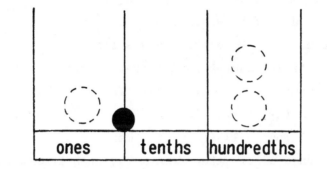

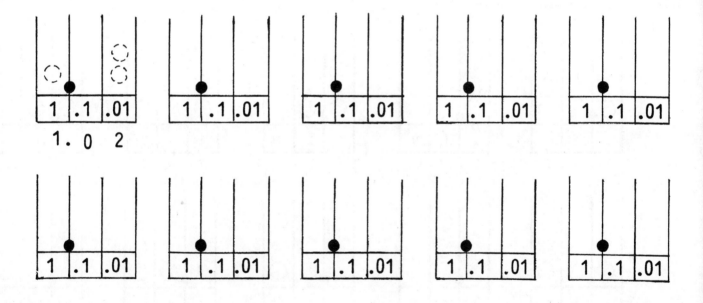

1. 0 2

How many different decimals did you find? _____

Arrange the decimals in order from largest to smallest.

Decimal Frames

Problem-solving skills pupils <u>might</u> use:

   . Record solution possibilities systematically.

   . Make a systematic list.

   . Look for patterns.

Comments and suggestions:

. This challenge should be used during the study of decimals or adapted as a whole number activity by using place values of hundreds, tens and ones.

1	.1	.01
3	0	0
2	1	0
2	0	1
1	2	0
1	1	1
1	0	2
0	3	0
0	2	1
0	1	2
0	0	3

. A systematic list, either by drawing or by a table assures pupils that they have found all solutions. The solutions (after the example) show first all possibilities with 3 markers in the ones place, then 2 markers, then 1 marker, then no markers in the ones place. With no markers in the ones place, the pattern shows solutions with 3, 2, 1 and no markers in the tenths place.

Answers:

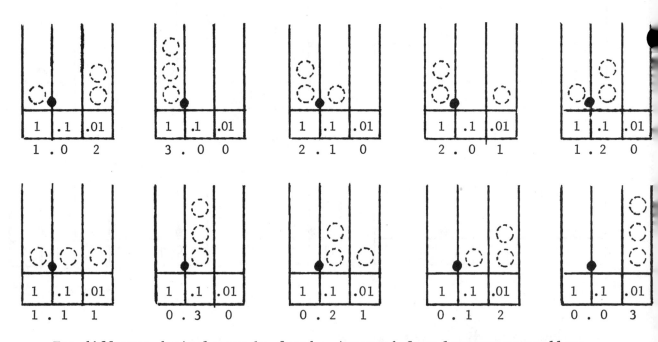

Ten different decimals can be found. Arranged from largest to smallest, they are 3.00, 2.10, 2.01, 1.20, 1.11, 1.02, 0.30, 0.21, 0.12, 0.03.

USING CLUES FOR SHAPES

Use the clues to decide which letter goes with which shape.
Write the letter on the shape. Some problems may have more
than one correct answer.

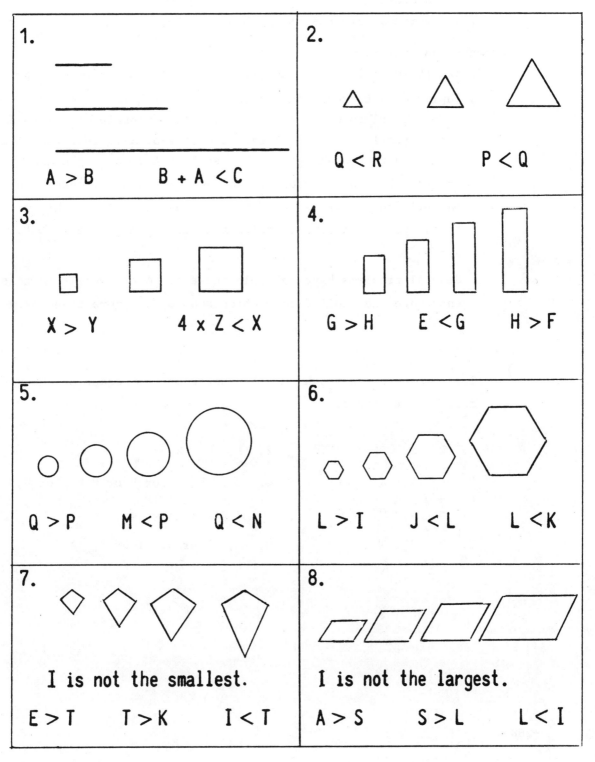

1.

A > B   B + A < C

2.

Q < R   P < Q

3.

X > Y   4 x Z < X

4.

G > H   E < G   H > F

5.

Q > P   M < P   Q < N

6.

L > I   J < L   L < K

7.

I is not the smallest.

E > T   T > K   I < T

8.

I is not the largest.

A > S   S > L   L < I

<u>Using</u> <u>Clues</u> <u>For</u> <u>Shapes</u>

Problem-solving skills pupils <u>might</u> use:

. Guess and check.

. Eliminate possibilities.

. Use a drawing.

. Recognize attributes of an object.

Comments and suggestions:

. Except for (1), the $<$ and $>$ refer to the areas of the shapes.

. Pupils need to know the meaning of the symbols $<$ and $>$. It is also helpful to understand that the relationship can be read in both directions. A $<$ B means the area of A is less than the area of B. Another meaning for this same statement is the area of B is greater than the area of A. In fact, for some of these problems it may help to rewrite some statements so each relationship reads the same. For example, in problem 5 a rewrite might be Q $>$ P, P $>$ M and N $>$ Q .

. Several problems have multiple answers. Alert pupils to this and encourage them to indicate which shapes have more than one answer.

Answers:

These answers are shown for the shapes as they occur on the page-- from left to right.

1. B, A, C

2. P, Q, R

3. Z, Y, X

4. E or F, E or F or H, E or H, G

5. M, P, Q, N

6. I or J, I or J, L, K

7. K, I, T, E

8. L, I or S, I or S, A

A MAGIC SOLID

Use the numbers 1-27 only once. Place them in the circles. The sum of any three connected circles is 42. Two rows of circles are connected by dotted lines to make the diagram less confusing.

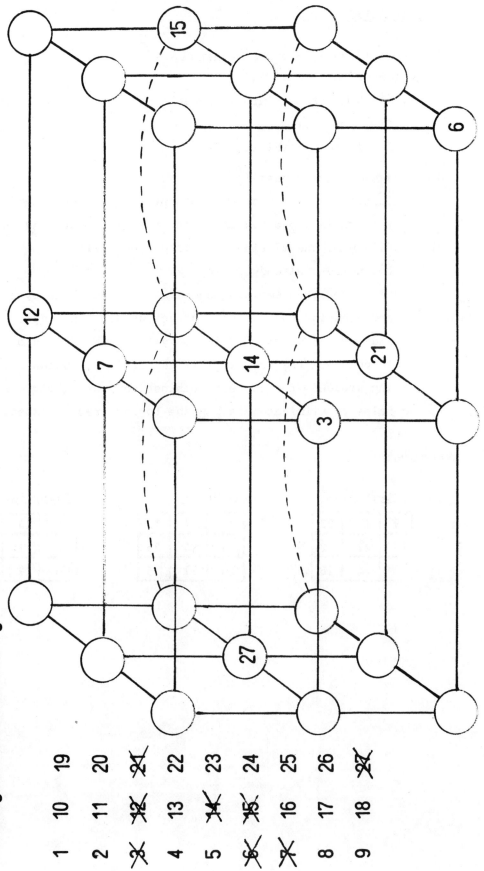

1   10   19
2   11   20
3̶   1̶2̶   2̶1̶
4   13   22
5   1̶4̶   23
6̶   1̶5̶   24
7̶   1̶6̶   25
8   17   26
9   18   2̶7̶

PSM 81

A Magic Solid

Problem-solving skills pupils <u>might</u> use:

. Interpret a drawing.

. Break a problem into manageable parts.

. Guess and check.

. Eliminate possibilities.

Comments and suggestions:

. Pupils may have problems interpreting the drawing. Perhaps relating the drawing to a box or to several blocks stacked together will help.

. Eleven of the missing numbers can be filled in by finding rows with two numbers already given.

. Clues for the remaining numbers are found both by studying the drawing and studying the numbers not yet used. The middle number on the right side of the drawing is 1. Only two combinations of numbers add to 41 to complete the row, 24 + 17 and 22 + 19. Guessing, checking and refining produces the correct combination. Using markers with the numbers helps to eliminate erasing due to incorrect guesses.

Answers:

Left Side

9	11	22
13	27	2
20	4	18

Middle

23	7	12
3	14	25
16	21	5

Right Side

10	24	8
26	1	15
6	17	19

THE RACE OF THE THREE STOOGES

Larry, Curly, and Moe decided to race up the stairs which have 24 steps.

Larry takes the steps 2 at a time while
Curly takes the steps 3 at a time while
Moe takes the steps 4 at a time.

1. If all three start at the bottom at the same time, who will

   finish 1st? _____ ; 2nd? _____ ; 3rd? _____

2. Where will Larry place if he has a headstart and begins on

   a. step 8? _____   b. step 14 _____

   c. step 12? _____

3. If Moe starts at the bottom, what
   step should Larry and Curly start
   on so all three finish in a
   tie?

   Larry _____

   Curly _____

-271-

PSM 81

<u>The</u> <u>Race</u> <u>of</u> <u>the</u> <u>Three</u> <u>Stooges</u>

Problem-solving skills pupils <u>might</u> use:
- . Use a drawing.
- . Work backwards.

Comments and suggestions:

. Emphasize that the strides taken by Larry, Curly and Moe take the same amount of time.

. Students might try a similar problem with real stairs or markers could be used with a larger drawing to show where each person is after each stride.

Answers:

1. Moe takes 6 strides to get to the top; Curly, 8 strides; Larry, 12 strides. Moe finishes 1st; Curly, 2nd; Larry, 3rd.

2. a. 8 strides needed--tie for 2nd with Curly.
   b. 5 strides needed--1st place.
   c. 6 strides needed--tie for 1st with Moe.

3. Larry needs a 12-step headstart. Curly needs a 6-step headstart. Pupils might work this backwards, all starting at the top and making strides downstairs until Moe gets to the bottom.

MAKING CHANGE

1. How many different ways can you make change for a quarter?
    Find out.
    Record your findings.   Compare results.

2. How many different ways can you make change for a dollar
    using 7 or more dimes and also using nickels and pennies?

Making Change

Problem-solving skills pupils <u>might</u> use:

 . Make a systematic list.

 . Identify patterns suggested by data in lists.

Comments and suggestions:

 . Allow students the freedom to decide their own methods for organizing a list. The answers given below are only suggestions.

 . The systematic list for problem 1 shown below keys on the number of dimes. It starts by listing solutions using two dimes, then 1 dime, then 0 dimes. Notice the pattern to the number of each--2 ways for 2 dimes, 4 ways for 1 dime, and 6 ways for 0 dimes.

 . The list in problem (2) keys on the dimes also. Notice the pattern-- 1 way for 10 dimes, 3 ways for 9 dimes, 5 ways for 8 dimes, 7 ways for 7 dimes. This pattern could be extended to arrive at 121 ways to make change for a dollar using dimes, nickels and pennies.

Answers:

1. 12 different ways

10	5¢	1¢
2	1	0
2	0	5
1	3	0
1	2	5
1	1	10
1	0	15
0	5	0
0	4	5
0	3	10
0	2	15
0	1	20
0	0	25

2. 16 different ways

10¢	5¢	1¢
10	0	0
9	2	0
9	1	5
9	0	10
8	4	0
8	3	5
8	2	10
8	1	15
8	0	20
7	6	0
7	5	5
7	4	10
7	3	15
7	2	20
7	1	25
7	0	30

BLESSED BEDBUGS

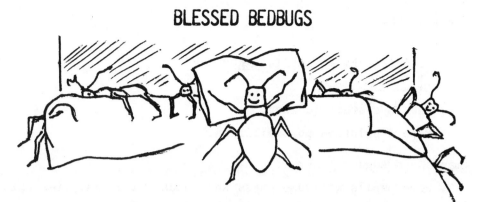

The Happy Holiday Hotel is blessed with cheerful bedbugs. Each single bed has 7 bedbugs. Each double bed has 13 bedbugs.

1. Suppose the hotel has 3 single beds.. Suppose there are less than 10 double beds. How many bedbugs could be in the hotel?

2. Suppose the hotel had these bedbugs. How many beds of each size would there be? If impossible, explain why.
   a. 20 bedbugs
   b. 40 bedbugs
   c. 140 bedbugs
   d. 75 bedbugs
   e. 45 bedbugs
   f. 115 bedbugs

3. The hotel added some king-size beds, and 5 bedbugs found their way to each one. The single beds still had 7 bedbugs each. The double beds had 13 each. Now the hotel has 45 bedbugs and 7 beds...How many beds of each size are in the hotel?

4. Create a bedbug problem of your own. See if a friend or your teacher can solve it.

Blessed Bedbugs

Problem-solving skills pupils <u>might</u> use:

. Guess and check.

. Make a systematic list.

. Record solution possibilities.

Comments and suggestions:

. Some pupils will use guess and check. They may view this strategy as saving more time than having to make a systematic list.

. Listing the multiples of 7 and 13 helps pick out those combinations which are solutions. Students might invent other ways.

Beds	1	2	3	4	5	6	7	8	9	10
Single	7	14	21	28	35	42	49	56	63	70
Double	13	26	39	52	65	78	91	104	117	130

Answers:

1. Possible numbers of bed bugs are 34, 47, 60, 73, 86, 99, 112, 125, or 138.

2. a. 1 single and 1 double.

    b. 2 single and 2 double.

    c. 7 single and 7 double; or 20 singles.

    d. 7 single and 2 double.

    e. Impossible. No multiples of 7 and 13 add to 45.

    f. 9 single and 4 double.

3. 1 single, 1 double, and 5 king size.